FORSAKEN ANGELS: IN THEIR OWN WORDS

THE PERSONNEL OF THE CHICAGO / NORTHWESTERN UNIVERSITY UNIT, U. S. ARMY BASE HOSPITAL 12, WORLD WAR I, 1916-1919

Edited by Michael W. R. Davis

FORSAKEN ANGELS: IN THEIR OWN WORDS
The Personnel of the Chicago/Northwestern University Unit, U. S. Army Base Hospital 12, World War I, 1916-1919

Edited by Michael W. R. Davis

Readers are encouraged to go to www.MissionPointPress.com to contact the author or to find information on how to buy this book in bulk at a discounted rate.

MISSION POINT PRESS

Published by Mission Point Press
2554 Chandler Lake Rd.
Traverse City, MI 49686
(231) 421-9513
www.MissionPointPress.com

ISBN: 978-1-943995-27-1

LOC: 2017905330

Printed in the United States of America.

"Military custom regarding nurses is most irrational. They are made officers and therefore not permitted to associate with enlisted men. This means that they must find their social life among other officers. But most male officers are married, especially in the medical corps. And most unmarried officers are from social levels into which nurses from small towns do not normally marry. As a result of this involved social system, military nurses frequently have unhappy emotional experiences. Cut off by law from fraternizing with those men who would like to marry them and who would have married them in civilian life, they find their friendships restricted to men who are surprisingly often married or who are social snobs."

— James A. Michener, *Tales of the South Pacific*, pp. 46-47. © 1946, 1947, 1960, The Macmillan Company, New York.

University Hall, Northwestern University.

TABLE OF CONTENTS

THE SECOND YEAR

May 18, 1918 to May 18, 1919

X APPENDICES

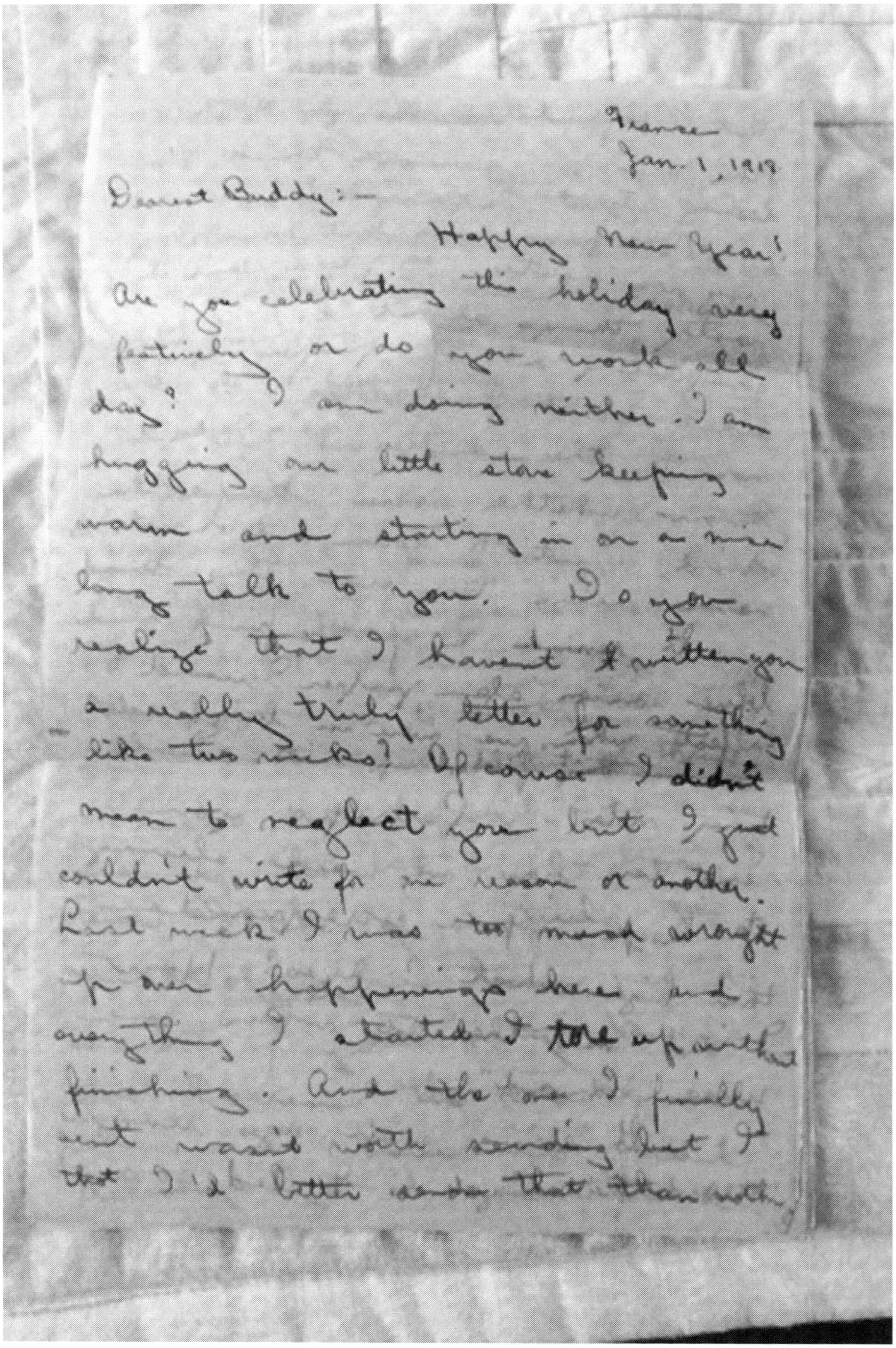
France
Jan. 1, 1919

Dearest Buddy:—

Happy New Year!

Are you celebrating this holiday very festively or do you work all day? I am doing neither. I am hugging our little stove keeping warm and starting in on a nice long talk to you. Do you realize that I haven't written you a really truly letter for something like two weeks? Of course I didn't mean to neglect you but I just couldn't write for one reason or another. Last week I was too much wrought up over happenings here and everything I started I tore up without finishing. And the one I finally sent wasn't worth sending but I thought I'd better send that than nothing

PREFACE

One day forty years ago, about ten years after our mother had died, my sister Ann Davis Skubik called me at my office in Dearborn, Michigan, from her home in Northern Virginia. "Did you know Mother kept a diary?" she asked.

"No, what was it about?"

"When she was a nurse in the First World War."

"Where'd you find it?"

"Well, you know I have her dresser." I recalled the beautiful light-wooded, gracefully bowed chest with the fourth drawer a secret one at the bottom. "Yes?"

"The drawers have been sticking, so I finally got around to taking everything out so I could fix it, and found this diary stuck in the back."

"Can you make a copy for me?" This was before the day of copy machines, not to mention home scanners.

Then I called my father who lived near us in a Detroit suburb. "Dad, did you know Mother kept a diary?" I asked.

"No, what's this all about?"

"Ann called me. She found it caught in the back of Mother's old dresser. It's about when she went overseas."

"No, I never knew anything about that. But I have all her letters, the ones she sent me when she went to France. If you'd like to have them, come on over."

He'd saved the letters, neatly tied with cotton strings in four small bundles, and packed in a grapefruit shipping box along with snapshots, picture post cards from his own duty in France with the American Expeditionary Force (A.E.F.), and yellowed newspaper clippings about my mother's hospital unit.

I sorted all these materials he had kept into piles spread over the living room carpet and sampled the journey to the past they revealed--images of my mother and her adventures, which I knew about vaguely but had never absorbed intellectually. As a journalist I sensed there was a grand story here, but not one I had time to explore at the time. So the keepsakes, along with a copy of the diary that my sister had dutifully sent along, were bundled up and put away for many years. Telling the story of Laura and her sister nurses of Base Hospital 12, known as the Northwestern (University) unit, was never forgotten, just delayed.

During a business trip to London a few years later, I took a few days off to trace Laura's footsteps from Falmouth to London, to Lady Henneker Heaton's country estate near Ipswich where she had been a guest, and finally across the English Channel to the French coast to locate the hospital site where, evading German bombings and suffering other hardships, she treated British and American soldiers for more than a year and a half.

Late in 2002 I finally undertook transcribing Laura Huckleberry's letters to John Erle Davis. Whatever happened to his letters to her is simply unknown. But her letters raised questions–such as how it was that civilian Red Cross nurses walked through a door and became U. S. Army nurses with no indoctrination whatsoever. This required research in the archives of Northwestern University, the Chicago Historical Society, the American Red Cross and the National Archives. These sources provided corroboration and perspective to Laura's diary and letters, which have been woven into this narrative.

In the main body of the text, commentary and interpretations by the editor are presented in italics while explanations are bracketed without italics.

Michael W. R. Davis

Royal Oak, Michigan

April 2012

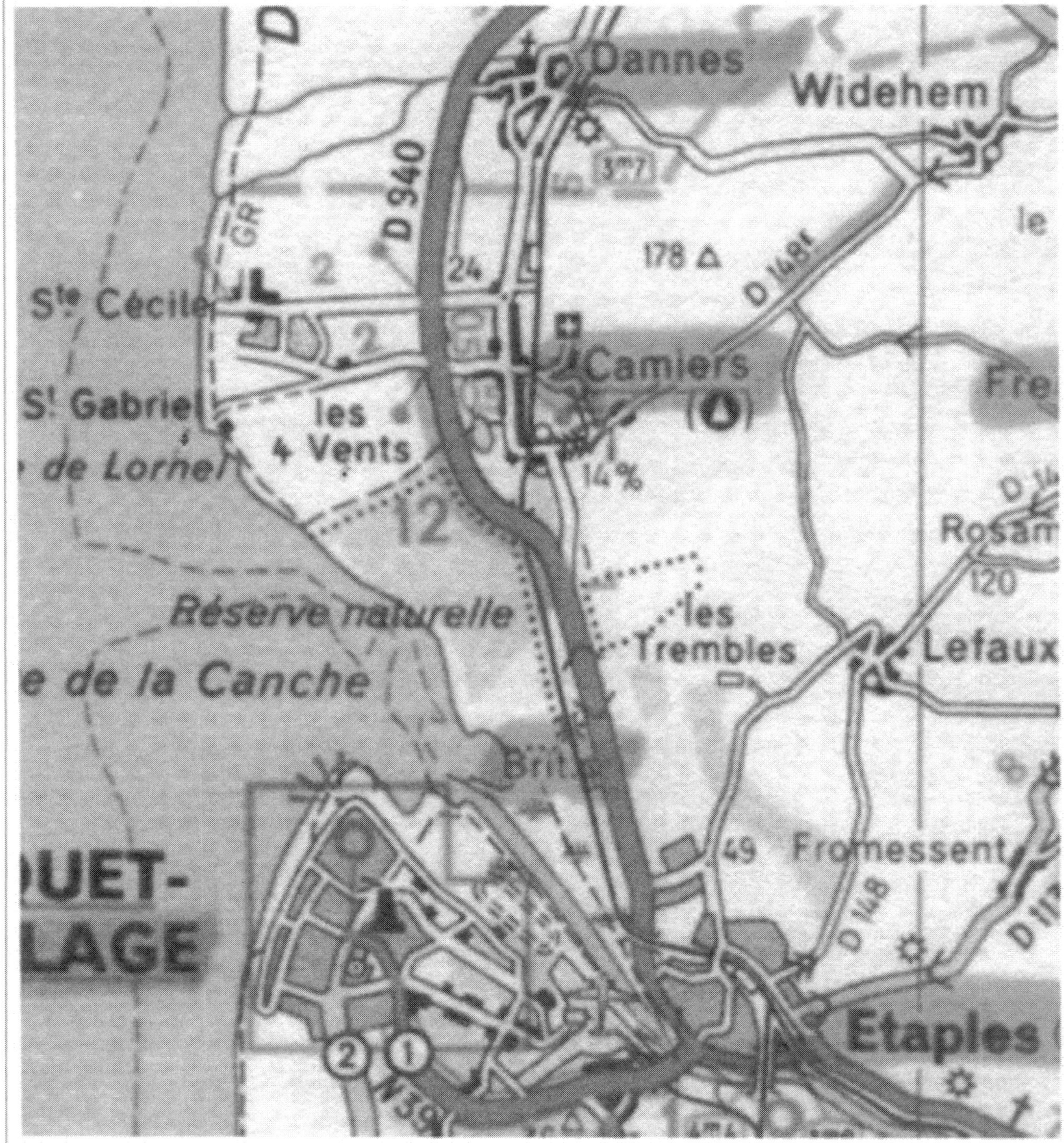
Dannes
Widehem
D 940
GR
24
178
D 148
St. Cécile
Camiers
St. Gabriel
les
4 Vents
de Lorhel
14%
12
Rosam
120
Réserve naturelle
les
Trembles
Lefaux
de la Canche
Brit
49
Fromessent
UET-
LAGE
D 148
Etaples
N 39

I. HISTORICAL INTRODUCTION

Nearly a century after what was variously known as The Great War, The World War and--after 1939--World War I or the First World War, first-hand memory of the 1914-1918 conflict is long-gone.

Because of that, it may be helpful to put that conflict in perspective for 21st century readers. Origins of The Great War date back to nation formation, particularly that of Germany, in the 19th Century, followed by a rush of the new major powers--not just European but the United States in North America and Japan in the Far East--to join old empires in building networks of overseas lands. It was also a time when nations formed alliances with other nations, ostensibly for defensive purposes. And it was a period in which centuries-old monarchies were gradually giving way to representative democracies or being threatened with the rise of totalitarian dictatorships. Some historians argue that the key influence was simply newly found national pride superceding centuries of religious, regional and minor principality identification.

Although the immediate trigger for World War I was the assassination of the Austro-Hungarian Empire's crown prince, the network of alliances actually forced the conflict. The chain went like this: Austria made demands on Serbia, where the murder of its monarch-to-be took place. But Serbia was formally allied with Czarist Russia, in turn allied with France. France was allied with Great Britain. The Kaiser's Imperial Germany was allied with Austria-Hungary and Turkey's Ottoman Empire. When Austria attacked Serbia, Russia attacked the Austro-Hungarian Empire and Germany came to Austria's defense against Russia. This brought France and Britain, the Allies, into the conflict against the Germanic Central Powers and Turkey. Italy and Japan joined the Allies. The United States and many peripheral European states including Scandinavia, Holland, Switzerland and Spain stayed neutral.

All the powers thought the conflict would be over in short order, but in a few months it had spread from the Balkans to Belgium and France, across Poland, the Middle East, and to where there were German colonies in East and South Africa, China and Pacific islands. It was truly a World War. Moreover, modern armaments, especially the machine gun, obliterated traditional means of fighting like cavalry charges and mass infantry attacks. On the Western Front, across northern France and the Belgian border, the war bogged down to hopeless, bloody trench warfare. On the high seas, the potent utilization of another new weapon of war, the submarine, eventually drew America into the war after American civilian lives were lost when the Germans sank merchant ships. The Americans arrived in Europe just as the Soviet revolution in Russia enabled Germany to concentrate all its forces against Britain and France.

In recent years, historians have revived considerable literary and scholarly interest in this early 20th Century war. Most recently, some scholars and historical journalists acknowledge that the seeds of the West's continuing conflicts in the Middle East were sown after World War I in the forced dissolution of the Islamic Ottoman Empire.

This account of Northwestern University's U. S. Army Base Hospital 12 in the First World War is important history because it was arguably the most distinguished such American outfit in the "War to End All Wars." The Northwestern unit was overseas longer than any other American unit of any kind, suffered the first casualties and treated more patients in France--over 60,000--than any other American hospital, with the possible exception of its neighboring Harvard Unit.

Moreover, this war became America's first adventure to the Old World, the land of their forefathers, for hundreds of thousands of energized, proudly democratic ordinary Americans. Cultural differences seemed inevitably to clash but instead, as the distinctly modern Chicago personnel of the Northwestern Unit found, those distinctions instead were accommodated.

In this book, Nurse Laura Huckleberry's saga, told through the initial diary and some 155 letters to her sweetheart, Chicago journalist John Erle Davis, has been augmented by official reports, newspaper stories, correspondence, and contemporary accounts and memoirs from other Unit 12 members. Laura's letters form the core of the narrative, revealing a gentle, suspenseful, true-to-the-era love story. Reminiscences of pre-war activities in Chicago and small-town Indiana provide a window into the social life of the times.

One unusual facet of this collection is that it was the man who saved the woman's letters. Conventional wisdom long held that only sentimental women saved such correspondence. Unfortunately, none of Erle's letters to Laura survive, although a few of his war letters to his own family are included. Today we enjoy instant, real-time communications across continents and around the world by Internet and email; but during World War I and until the advent of airmail, a complete exchange of letters—and answers to questions--took six to eight weeks or longer for forwarded mail, by ship and train.

Base Hospital 12 was Chicago's first substantial contribution to the First World War. By 1917 Chicago--incorporated only 80-odd years before—had become America's second largest city with a population of over two million, the center of the Midwest's commerce, culture and transportation, and truly the capital of America's Inland Empire. With the opening of the Welland Canal between Lakes Ontario and Erie in 1881, Chicago also had become a world port for ships able to fit in the canal's 270-foot length. But until the World War, Midwesterners were far less likely to travel to Europe than those on America's Atlantic Coast.

The memoirs of Pvt. George R. Baker—edited and published by his son George R. Baker Jr. in 1999 under the title *Heroes and Angels*—are especially noteworthy. Not being created under censorship, they portray an entirely different and often harrowing experience than that which Laura was able to relate in her letters, which had to pass under the censor's eye. Excerpts from Pvt. Baker's memoir included in this volume also provide the perspective of both an enlisted man and a "university man"—and, to boot, an avid athlete.

Finally, the chronological progression of these letters, memoirs and other documents also provides an unexpected window to the typical American's ready adaptation to other's languages and expressions--in the preserved, often contemporary words of those who were there. Over the months, "trucks" osmosed to "lorries" and "very good" became "tres bien."

II. BIOGRAPHICAL NOTES

Laura Gertrude Huckleberry was born on the 11th of January during the Great Blizzard of 1888, at the rural Jennings County, Indiana, farm established by her grandfather in 1834.

The family name Huckleberry was an Anglicization of the German name Hagelberger. Laura's ancestor Benjamin Hagelberger had emigrated with his children to Pennsylvania in 1752 from the Palatine village of Rott on the west bank of the Rhine River near Strasbourg, now part of the French province of Alsace.

Laura's other forbears were Scots-Irish, Welsh and English Protestants who'd migrated from Pennsylvania, Maryland and Virginia to the middle Ohio valley either down the river or across the Appalachian mountains through Kentucky. Her father, a Union veteran like Laura's grandfather, had been a farmer, schoolteacher, deputy sheriff, county clerk, and was now a rural mail carrier. Her maternal grandparents had migrated to Indiana, also in the 1830s, from Tennessee and Kentucky. She was the youngest of five, her siblings being brothers Warren, Will and Silas, and sister Bertha.

Laura graduated from North Vernon High School in 1906 and spent the next two years teaching at a rural, one-room schoolhouse to save enough money to attend Indiana University for a year. Then she heard about nursing, the new profession for women. A professional nursing education was available through the three-year co-op program at the Illinois Training School for Nurses, associated with Cook County Hospital in Chicago. She had been raised as a Methodist, an important factor in this narrative. At the time of her enlistment in the Red Cross Reserve in early 1917, Laura was a nurse at Children's Isolation Hospital, also associated with Cook County Hospital.

John Erle Davis, the recipient of Laura's letters recorded in this narrative, was also 29 years old when the U.S. declared war April 6, 1917. Like Laura, his roots were deep in Jennings County, Indiana. His great-grandfather Phanuel Davis, who'd served as an Ohio Militia sergeant in Detroit in the War of 1812, had established a farm in the county's Campbell Township by 1820. Phanuel's wife was descended from English, Scottish and Protestant-Irish stock who were among the earliest pioneers on the Virginia frontier in the 1700s. Erle's Presbyterian grandfather married an Irish immigrant in 1843, and thereafter that branch of the family was Roman Catholic.

The Davis and Huckleberry farms at one time adjoined at one corner, but in rural Indiana a century ago, Protestants and Catholics scarcely spoke to one another, let alone socialized. As Laura's wartime letters reveal, she and Erle had been interested in one another since high school days, a relationship growing when they both worked in Chicago in the immediate pre-war years.

Erle's father had been a stonemason and foreman of stone-worker crews building roadbeds and bridges for the Baltimore & Ohio Railroad. In 1917, was a clerk in the State government in Indianapolis. Erle was the oldest of three, his siblings being sisters Audrey Elizabeth (Betsy) and Harriet (Hettie). Because he had lived for years with his aunt, Mrs. John N. Rees, her sons Jack and Billy, midshipmen at the U. S. Naval Academy during the war, were like brothers to him.

In 1917, Erle was a journalist who'd dropped out of high school in the 11th grade to learn the printing trade, later becoming a reporter and editor in Louisville, Kansas City and Chicago. At the time his Illinois State Guard unit was federalized in May, 1917, he was an editor with the Western Newspaper Union of Chicago, a syndicated news service for weekly newspapers.

OTHER KEY PERSONS IN THE NARRATIVE, ARRANGED ALPHABETICALLY:

(Biographical sketches may be found in the Appendix)

Ayres, Mrs. Edith, *Reserve Nurse, U. S. Army Base Hospital (BH 12), one of Laura's Chicago roommates; all of the Chicago roommates as well as Miss Urch and Miss Spencer were graduates of the Illinois Training School for Nurses*

Baker, George R., *Private, BH 12*

Besley, Frederick A., M.D., *Reserve Major who organized BH 12*

Cleveland, Elizabeth T., *Reserve Nurse, BH 12, Laura's roommate in France*

Collins, Christopher C., *Major, U. S. Army, BH 12's original commanding officer*

Connard, May (Conny), *Reserve Nurse, BH 12 replacement, one of Laura's Chicago roommates*

Dancey, Jesse S., *The Rev.*, *Methodist Chaplain of BH 12*

Gambee, Bess, *Reserve Nurse, BH 12, one of Laura's Chicago roommates*

Glaspel, Cyril. J., M.D. *Lieutenant, BH 12*

Hall, Arthur R., *Private, BH 12, former Northwestern University law student*

Hampton, Frances B., *Reserve Nurse, BH 12, one of Laura's Chicago roommates*

Horowitz, Gabrielle, *Reserve Nurse, BH 12 replacement*

Lyon, Elizabeth C., *Reserve Nurse, BH 12*

Krost, Gerard N., M.D., *First Lieutenant, BH 12*

Krost, Carrie G., *Reserve Nurse, BH 12, Dr. Krost's wife*

Mandel, Milton, M.D., *Captain, BH 12*

Matzen, Emma, *Reserve Nurse, BH 12, one of Laura's Chicago roommates*

McCosh, Duncan, *Lieutenant, Canadian Army*

McMillin, Edith (Mac), *Reserve Nurse, BH 12, one of Laura's Chicago roommates*

Murray, Robert, *Lieutenant, His Majesty's Imperial Army*

Nadler, Walter H., M.D., *Captain, BH 12*

Neal. M. Pinson, M.D., *First Lieutenant, BH 12*

Nussbaum, Payson L., M.D., *Captain, BH 12*

Pawlisch, Ella R. (Pawley or Polly), *Reserve Nurse, BH 12, one of Laura's Chicago roommates*

Speed, Kellogg, M. D., *Captain, BH 12*

Spencer, Ruth, *Reserve Nurse, BH 12*

Urch, Daisy Dean, *Chief Nurse, BH 12*

All the Northwestern Unit's original personnel were native-born Americans except for Moses Blumenthal, the pharmacist, born in Russia.

III. THE BACK STORY

U.S. DECLARATION OF WAR AND BASE HOSPITALS

When the United States declared war on Germany on April 9, 1917, there were only 403 nurses on active duty, including 170 reserve nurses who had been ordered to duty in twelve Army hospitals in Texas, Arizona, and New Mexico as a result of incidents on the Mexican border. By the end of June 1917, there were 1,176 nurses on duty. One year later, 12,186 nurses were on active duty serving at 198 stations worldwide, recruited almost entirely from the ranks of "reserve" American Red Cross nurses.

Immediately after America's declaration of war in 1917, British representatives rushed to Washington to explain that the Allied war effort's most pressing need was for hospital services in France, to relieve British units exhausted by two-and-a-half years of brutal casualties. Thus in May 1917 America dispatched six base (general) hospitals to France with more than four hundred nurses.

To quickly complete Chicago's Base Hospital 12 staff, augmenting the volunteers who had enlisted before America entered the war, Dr. Frederic A. Besley of Northwestern University Medical School relied on professional colleagues from the school, nurses and other personnel from Cook County and Evanston hospitals, and young enlisted men principally recruited from the Chicago area's college campuses or from the ranks of recent graduates.

When Base Hospital (BH) 12 departed from Chicago on May 16, 1917, it consisted of 26 officers, 65 nurses, four civilians and 153 enlisted personnel--including a small cadre of Regular Army officers and non-commissioned officers. They were joined later by reinforcements. A number of the original personnel were promoted or transferred to other units as America's contribution to the war effort built up.

In addition to what came to be called the Northwestern unit, other pathfinder base hospitals sailing for France for service with British Expeditionary Forces included Presbyterian Hospital (BH 2) from New York City, Lakeside (BH 4) from Cleveland, "Harvard Unit" (BH 5) from Boston, Philadelphia (BH 10) from Pennsylvania Hospital, and Washington University Medical School (BH 21) from St. Louis. Together, these were the first major U.S. Army units sent overseas in the First World War.

In the case of BH 12, the Chicago branch of the Red Cross paid for supplies, equipment and some salaries when the unit went overseas. In 1916, the Chicago branch raised $243,000 to supply and equip military hospitals, later allocating $25,000 to equip BH 12. In preparation for 12's departure for France, on May 3, 1917 the chapter pledged $3,000 per month for salaries and supplies for the chaplain, dieticians and other civilian support personnel. And in August 1917, the Chicago Red Cross expended $6,000 for uniforms and equipment contracted for before the unit embarked, and later appropriated $30,000 for the Northwestern Unit's X-ray, surgical, kitchen and laundry equipment, mess gear, a truck and an ambulance. In December, Chicago sent $3,000 for a portable electric generator to power BH 12's X-ray equipment.

The first indication that American nurses were forsaken came with their

THE NURSES HOME. (1) ORIGINAL BUILDING OF 1883, WITH FRONT AND FOURTH STORY ADDED IN 1887. (2) CRERAR ADDITION, 1907. (3) MACDONALD PROPERTIES, PURCHASED IN 1910

official status and pay. According to a September 4, 1917, report by E. Percy Noel, *Chicago Daily News* staff war correspondent, "The Canadians give army nurses the rank of captain, entitling them to a salute from the men." In 1917, a U. S. Army captain on duty overseas was paid a mere $200 a month. But the Army paid an American Red Cross/U.S. Army Reserve Nurse, who had officer status but no commission, only $60 a month while overseas--the equivalent of slightly over $1,048 a month in 2011, or $12,576 annually. As an indication of progress, today's Army nurses with training and experience comparable to most of those in 1917's Base Hospital 12 earn about $60,000 annually in base pay plus overseas, hazardous duty and other allowances.

THE FIRST YEAR

MAY 17, 1917 TO MAY 15, 1918

Diaries, memoirs, news reports and letters

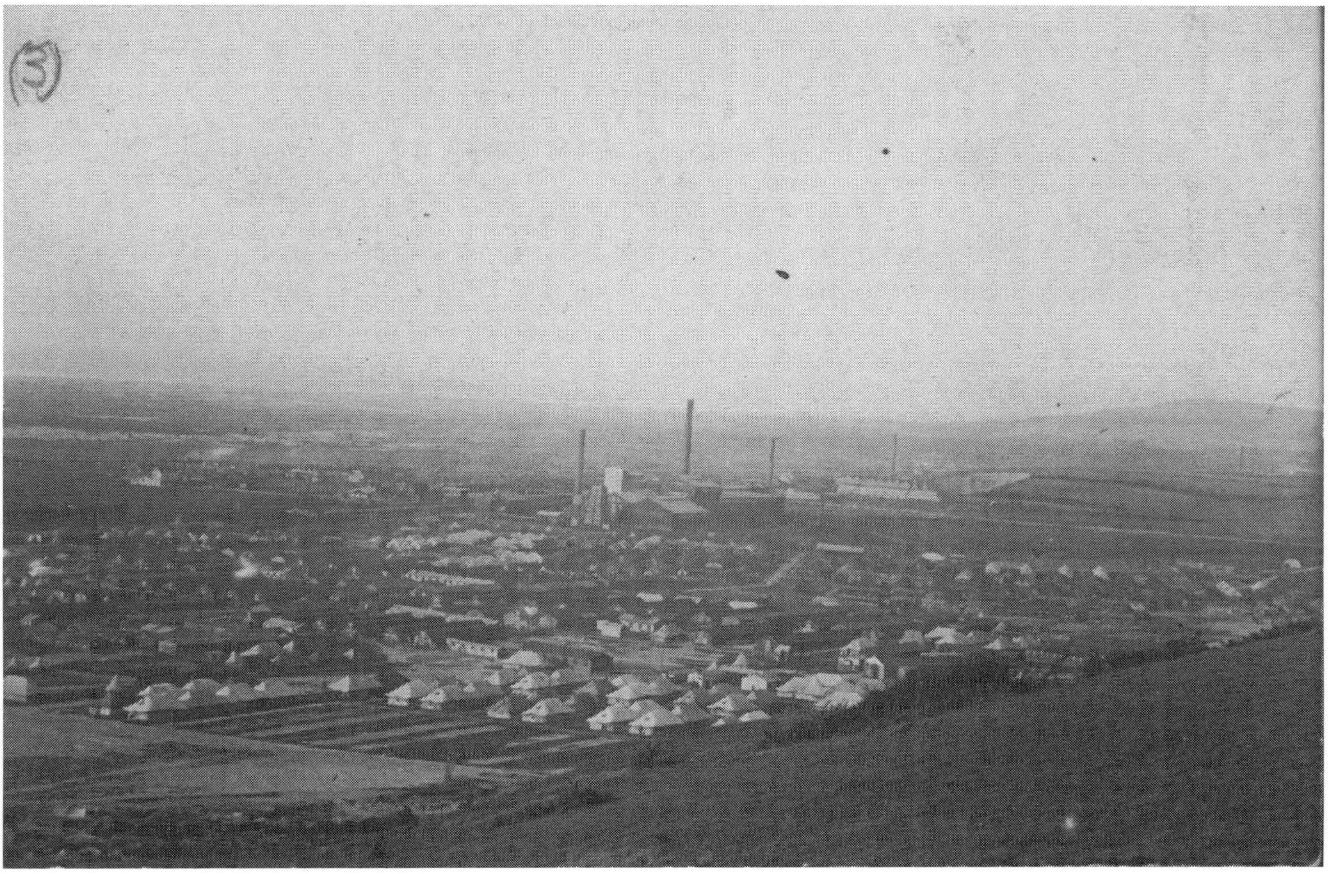

The complex of military hospitals at Dannes-Camier, France, with the English Channel in the distance. Here, U.S. Base Hospital #12 took over British General Hospital #18 in June 1917 for the rest of the war.

IV CALLED TO DUTY

*The Chicago Herald, Tuesday, May 2,1917**
HOSPITAL UNIT GETS ORDERS

The Northwestern University hospital unit No. 12, which was organized under the direction of the Red Cross several months ago, yesterday received orders from Surgeon General Gorgas of the United States army to prepare to leave Chicago for the east at a moment's notice. Twenty-three doctors, two dentists and fifty graduate nurses have already been sworn into service of the United States army. The remainder of the 196 which make up a complete unit will take the oath immediately, according to Dr. Frederic A. Besley, who is the director of the unit. The nurses' aides, orderlies, sergeants, and privates have been notified and are making ready for duty. Dr. Besley's personal staff consists of Dr. Kellogg Speed, surgeon; Dr. Milton Mandel, chief medical officer, and Miss Daisy Urch, chief nurse . . .

It is expected that the unit will be ordered to leave for France within a very short time to take charge of a base hospital containing 500 beds.

One purpose in sending the unit abroad is to enable the doctors to study the newest methods of treating wounds and diseases for the benefit of the prospective American expeditionary force.

* *Many of the newspaper stories reproduced in this narrative were found in a Base Hospital 12 scrapbook at the Northwestern University archives, randomly pasted into available space and unidentified as to particular newspaper or date published.*

The Chicago Herald, Thursday, May 17, 1917
CHICAGO UNIT OFF FOR FRANCE
Base Hospital No. 12 Leaves for Eastern Port, Bound for the Front

Bound for "somewhere in France," its exact destination unknown even to its commanding officer, Chicago's first army unit—base hospital No. 12—left last night for an eastern port. It will embark for Europe within ten days.

Tears, laughter and a shouted chorus of good-bys marked the departure of the special train. There was no music and no public demonstration of any kind. The train's route, time of departure and expected arrival are withheld by the *Herald* at the request of Dr. Frederic A. Besley, director of the unit.

The personnel of the unit comprise 247 persons, of whom twenty-four are physicians, sixty-five trained nurses, two dentists and 153 enlisted men*. Seventy-five per cent of the enlisted men are college students, about one-third being from Northwestern University.

Just before the train departed Dr. Besley secured long-distance telephonic communication with his mother, Mrs. W. B. Besley, at San Diego, Cal. Her answer to her son's announcement that he was leaving for Europe was:

"I'm awfully sorry, Fred, that you're going, but I'd be sorrier still if you weren't."

Major C. C. Collins and Captain A. E. Magee of the U. S. Army are in military charge of the unit. Captain John A. Porter, U.S.A., will act as quartermaster.

* *There are discrepancies from one source to another, even in "official" rosters, as to the number and distribution of BH 12 personnel. Beginning and ending rosters are included in the appendix. In addition to Regular Army Major Collins and Captains Magee and Porter, there was a cadre of five regular Army sergeants whose task it would be to provide basic training in military matters for the unit, especially the 150 or so privates. Within a few months, several of the doctors had been promoted from their original commissions, 9 of the privates had been promoted to sergeant first class, 16 to sergeant, 9 to corporal and 45 to private first class. By the time they returned to Chicago in 1919, many more had been promoted in both officer and enlisted ranks.*

The Nurses

Excerpt from Memoir of Reserve Nurse Elizabeth Lyon, prepared in 1975 for alumnae of the Illinois Training School for Nurses:

In the Spring of 1916 there was much discussion about the proposed plan of organizing Base Hospitals under the Red Cross to be called into active service in case of war or other great need: the plan being to select doctors, nurses and orderlies from one hospital—people who were accustomed to working together, knew each other's ways and thus could do good team work.

It seemed that the great number of doctors and nurses who had training at Cook County Hospital should certainly furnish enough for at least one or two such units. About this time Dr. Frederick Besley, attending surgeon at Cook County Hospital asked Miss Daisy D. Urch if she would be willing to undertake the interesting task of organizing the nursing service for such a unit.

She began at once to stimulate enrollment in the Red Cross as a preliminary measure so that by February 1917 when authority to organize Base Hospital 12 was finally given, a large number of nurses were enrolled and some had even taken the required prophylactic smallpox and typhoid vaccine. However, people were loathe to believe that America would declare war and even if she did that soldiers would be sent overseas and that there would be many nurses needed. So it required considerable stimulation to arouse any enthusiasm. Still we kept steadily adding to our numbers . . . according to the plan outlined by the Red Cross. The nurses selected were either graduates of the Illinois Training School for Nurses or affiliated schools with about a dozen from Evanston Hospital. The officers were ex-interns from Cook County Hospital.

America's declaration of war was a salutary stimulant to get our unit in order but even then very few believed we would ever be called into service. However, on Monday morning May 1st, 1917, out of a clear sky came a telegram from Miss Noyes at Washington to mobilize at once for service in France followed by another with instructions to take sixty-five nurses instead of the original number, fifty. Some readjustments had to be made for we had nurses fitted for home duty but not for "overseas" duty.

On Wednesday evening, May 16th, 1917 the entire unit—officers 27, nurses 64, and men 147—assembled at Polk St. Station, took the oath of office and boarded a special train for New York.

The "University Men" in fact reported in civilian clothes with suitcases, musical instruments and sports equipment. Totally untrained in Army ways, they were not issued uniforms or equipment until they were on shipboard. Such was the poor state of American preparedness when it entered the First World War—as in other foreign wars over into the next century.

The University Men

Memoir of Private George Baker:

At the time America entered the war with Germany, I was attending the Young Men's Christian Association College in Chicago.

When President Wilson declared war, I investigated the various branches of services carefully before enlisting. I wanted to get into an outfit that would go overseas at once because we all thought that the war would last but two or three months.

About May 1, 1917, three weeks after the U.S.A. declared war, I heard of the hospital outfit at Northwestern University in Evanston, Illinois, which was ready to leave for France to join the British Army. I decided right then to try to get in.

I called up Chaplain Dancey, who was Pastor of the First Methodist Church in Englewood, Illinois—"my own church"— about enlisting in this special outfit. He gave me the recommendation and I went out to Evanston on May 8 and talked to doctors and some of the officers and they told me to come back next day, May 9. This I did and was examined and signed up for immediate service.

I then went home to Rock Island for a day or two and on Monday, May 14, 1917, the whole outfit took the oath of enlistment in the regular U. S. Army for three years, and four years in the Reserve. My unit was Base Hospital #12, Red Unit.

It all seemed so strange— one day in school and the next day in the Army ready for a 4,500 mile journey to get to a war that had gone on for two years and nine months already. We thought we would have a nice long trip and the war would be over, but we found out later we were mistaken.

Troop Train to New York

Reporting to duty in civvies.

Pvt. Baker's memoir:

At 8:00 p.m., Wednesday, May 16,1917, the 147 enlisted men in our unit marched away from Patten Gym at Northwestern University with our baggage, suitcases and all kinds of bundles, amid great shouting, singing, Rah! Rah's! and good-byes from the students. We all piled on the Evanston "L" and got off at Union Station.

We left Chicago at 11:00 p.m. on the Wabash Railroad on a special train that had regular sleepers and meals. In Canada, at St. Thomas and Hamilton, the people greeted us. We had written in chalk on the side of the cars, "BERLIN OR BUST". The Canadian people were doubtful about our getting to Berlin.

We arrived in New York at 5:30 a.m. on Friday, May 18, 1917 after thirty-one hours of steady riding. After being told not to tell anyone where we were going and not to send any information through the mails, we were given the whole day in New York.

Letter #1

Reserve Nurse Laura Huckleberry
to John Erle Davis:

On the train
Somewhere in Canada
Thursday 1:30 PM
5-17-17

Dear Buddy: —

I want to call you that, too—All right?? I dreamed and dreamed and dreamed about you last night—always walking along just touching your hand. I wonder if you dreamed also.

We were assigned our berths and I drew Upper 9. Frances is in upper 7 but Bess & Mac were fortunate enough to be together in upper & lower 6. The other girls are in the other cars but none alone.

...

Just at that moment we pulled into St. Thomas & such an ovation!! There was almost the whole populace down to greet us. They passed us Canadian & British flags*, ice cream cones, oranges, bananas and addresses galore. Sons, brothers & husbands to look up or look out for. One old gentleman told us they have sent 2000 soldiers & their population being only 19,000 you may see how much it meant to them. It was all quite wonderful but made swallowing a bit difficult for a while.

We don't know when we'll get to N.Y. nor when we leave there. If any one knows they keep it to themselves. But anything addressed to the Red Cross Headquarters at N.Y. will reach me. (Hint ??? Oh no, my dear!)

It's almost time for lunch so I'll stop now and see what is on the menu. Please excuse the scribble—it's partly due to the train's wriggling.

Try not to be lonely and write often and come soon.

Yours as ever

Laura

* *A tiny red Canadian flag on a hatpin was fastened to the page.*

Laura's first wartime letter to Erle was dispatched even before the troop train reached New York City. The envelope was marked in Laura's handwriting Special Delivery, postmarked Buffalo, NY, May 17 11 PM, and addressed to:

Mr. Jerle Davis
210 S. Desplaines St.
Western Newspaper Union
Chicago, Ill.

Letter from Erle to Laura's sister, Bertha Huckleberry, a schoolteacher in North Vernon, Indiana:

Editorial Rooms
Western Newspaper Union
Newspaper Service
210 So. Desplaines St.

Chicago, May 18, 1917

My dear Miss Huckleberry, —

Laura and her hospital unit left Chicago for France at 11 o'clock Wednesday night. She thought you might be interested in the newspaper accounts of the departure, so I am sending those which appeared yesterday morning.

As departures go, this event was an unintentional comedy of high quality. The unit was composed of about 350 persons, and think each had a separate following of, say, 500 friends, relatives and enemies on hand to perform the ceremonies of farewell. When all were assembled it made quite a considerable foregathering, and it sounded like the bird house at the zoo.

Too, there were among those present a number of kind old ladies who went thither and yon distributing to the brave Red Cross lassies little bunches of wilted flowers, small flags and books. Laura received her volume just as she squeezed through the gate to the train shed. It was really handsomely bound and led to great expectations. After she had reached the coach and found her berth, she turned eagerly to an examination of the tome. All was revealed in the title page; it said: "Rob Roy." Ever hear of the story? Neither did we. But you never can tell what kind old ladies will do.

One of my kinswomen sent me a clipping about myself which she had lifted from the North Vernon Plaindealer. It was certainly a monument of misinformation and a vast libel on my patriotism. I am not going any where as a special correspondent; but I hope to start for France soon with my hospital unit and I trust I shall be given an opportunity to perform as a private. I believe I'd do any thing to get into this fight. It is unlikely, though, that unit eleven, of which I am a member, will leave Chicago before the latter part of June.

Surely you will forgive the commercial aspect of this letter when I tell you I was unable to find a single sheet of polite stationery in the whole works.

Sincerely –

J Erle Davis

Letter #2

Laura to Erle:
The Waldorf-Astoria
New York, 5-18-1917

Dear Erle: —

We are being entertained by the Red Cross here and my respect for the Lady Boards of such things [is that they] are exceedingly important & so tiresome. They gave us dresses, hats, capes, brassards & caps and the poorest breakfast you ever saw.

We are to be here for lunch with them. We were all dubious of the invitation after our introductory meals but have decided to risk it for the sake of saying that we lunched at the Waldorf-Astoria.

So far we've seen but little of New York—just from the top of a bus on our way up from the station. We did come up 5th Ave. anyway.

Our capes are navy with bright red flannel lining & a red cross on the front—quite good looking. The hats are blue velour & soft & small & I'm much better pleased.

We have just received word that the Lakeside unit from Cleveland arrived all OK.

We sail tomorrow noon on the Mongolian or Mongolia—didn't quite catch the name but that's it anyway. Later this [***indicating Mongolia***] is correct

This afternoon if we ever get away we want to take an automobile & see what we can. We must sleep aboard ship tonight & will not be allowed off tomorrow so this is probably my last note till we're off.

The red tape takes so long but of course it must be put up with. The "privates" with us are all Northwestern University students and they've been singing & whistling & yelling. They hustled our trunks & suitcases aboard the ferry & off again. That's all the work I've seen them perform.

Good bye till I have a moment aboard ship to write again.

Laura

The red tape takes so long but of course it must be put up with. The "privates" with us are all Northwestern University students and they've been singing & whistling & yelling. They hustled our trunks & suitcases aboard the ferry & off again. That's all the work I've seen them perform.

Pvt. Baker's memoir:

Friday, May 18

We went to our quarters assigned on the S.S. Mongolia and watched it loaded. We were given second-class bunks with sheets and pillows. Gee! We went to war in style. We all wore civilian clothes so that any German spies, of which New York was full, who might be sneaking around, would not think we were Regular Army men. Well, we had our first meal on board ship and what a riot. Wow! Rotten meat, cold spuds, stale prunes and fish and butter so strong it would get up and walk along the table. We were dumbfounded at such a line-up and the thought of what we were in for.

The dining room was one deck below main deck, even with the water line and had been used for carrying cattle previously. Imagine a group of college men jumping from the luxury of peace into a cattle pen for the grub. There was only one entrance to our dining room and that was about ten feet down a straight ladder. Well, we sure crabbed and we howled so loud we got the ship's head steward and our officer-in-charge and demanded sanitary food. It ended by putting two of our men in the kitchen. The Government was paying so much per head—private contract—and that is the reason for the rotten grub. The boys threw the butter out a porthole at one end of the table after it had been passed along the whole length of the table as we held our noses. Wow-ie! The war had started at last. All the way over food was not much better even though the Government paid plenty for it.

Letter #3

Laura to Erle:
On board the Mongolia,
Sat. morning. [May 19]

Dear Man: —

We saw just a little of New York yesterday & Chicago really looks like a little country berg beside it. Honest!

I know you'll like Central Park & the Hudson & everything. I saw or rather rode past St. John's Cathedral—recognized it without being told—wonderful intelligence.

We are on a freighter that has been fixed up for a transport so it isn't much in the way of a boat but then you see, never having been used to floating palaces I'll be able to get along nicely. The ones who have crossed before say that the boat is so small it will probably roll about considerably but that also will be an experience.

Mail was just called & I got a nice fat one from you. Many thanks. Think I shan't open it till we start—just for fun & to finish this.

Be sure to bring your kodak. A number of the boys have them & so they'll probably get thru all right.

The trouble is that no one seems to know just what they want or can do.

They're waiting for this to go ashore

Much love

Laura

Dangerous Journey and Chicago's First War Casualties

Pvt. Baker's Memoir:

About 1:50 p.m. Sunday, after twenty-two hours out and about two hundred fifty miles from New York, we had target practice with our three 6" guns manned by twenty-two sailors and a Lieutenant from the Regular Navy. I was watching out my porthole part of the time. They shot at barrels which they had dropped overboard and then at half mile would try to hit them. Every shot would make the old ship quiver from stem to stern. The ship was six hundred feet long—about two city blocks—with a beam of ninety feet, a large merchant ship. One shell which was being fired from a forward gun, exploded opposite the ship about two hundred feet from me and killed two nurses and wounded another in the thigh with a large piece of shell case. We were just twenty-two hours out of New York so Captain Rice turned back to New York to leave the bodies. If we had been twenty-four hours out we would have kept on going.

Letter #4

Laura to Erle:
S.S. Mongolia
Sunday PM [May 20]

Dear Erle: —

We're on our way back to New York as a result of an accident during target practice. It makes me feel so small and insignificant in this world. We can never tell when the call will come and yet I have no fear.

We were all grouped on the upper deck watching and all happy and sort of carefree. We were having such glorious, warm weather and the water was so smooth and no one was sea sick.

Then some one said "Somebody's shot," I turned and saw Edith Ayres on her face on the deck and Emma Matzen holding her coat where a big hole had been torn. Edith had a temple wound which took her instantly. Miss Wood, an Evanston nurse, was shot in the heart & and Emma had only this flesh wound on the hip & on the arm. She is all O.K. unless infection starts in. How or why the rest of our little family* escaped, I don't know, for we were all together and the Evanston girls & a bunch of Drs. all in a bunch beside us.

We are to leave Emma in N.Y. so in case of being torpedoed she will not have to risk being carried out. Our happy family is now reduced to five. I hope that it will suffer no further reduction.

We were all more or less expecting to be torpedoed but this sort of took us off our feet—coming so soon & from such an unexpected source. Did I tell you that the shell was defective and exploded as soon as it left the gun? Or else the gun was defective. I've really been unable to ascertain the exact truth but think that the ammunition was to blame.

I don't know whether we'll just anchor in N.Y. harbor for awhile & then start again or if we'll be put on another transport. In fact I even don't know whether you'll ever get this. It (the mail) may all be destroyed. Masters of our fate?? Well hardly. But then you know we'd probably not do so well. We'd all expected something and we're just finding out what war is. . . .

We nurses go as first class passengers and have practically the same privileges as the officers but the privates aren't treated so. But of course you know that. . . .

The boys have not yet been given their uniforms. They're all such nice youngsters we did hate to see them sent down to the steerage when we started out. And they aren't even allowed on the upper decks and their food is very poor, they tell us. I've slipped several apples from the table to throw to them and one orange. Of course that doesn't fill any gap any place but it relieves my feelings a bit anyway. . . .

I wrote nothing of the accident home, thinking that by some chance it would not meet their eyes in the papers. But you are different, you'd probably see it if it gets into the papers at all. And if it doesn't get to the papers, you'll probably not get this

. . .

Miss Urch just came in & said that 2 of the girls would go ashore with Emma so I'm sure I can slip this in some way.

We were supposed to take about 10 days for the trip & now we'll be longer. I do so dread starting again. We were so happy when we left the docks yesterday afternoon. It has taken lots of pep but this may be only the means of saving others. I understand that the gun had not been tested out before.

My room mates have come in and as our state rooms are 2 x 2 I'll have to stop. . .

**Laura and her nurse roommates in Chicago*

The *Mongolia* Incident

The Mongolia incident—involving not just the first American casualties after war had been declared, but nurses as well—produced headlines and major front-page stories on Tuesday, May 22, 1917, all over the country. Against rules, Laura managed to slip her letter to Erle about the accident to her shipmate going ashore in New York, and veteran newsman Erle in turn leaked it to the Chicago Tribune *since the weekly syndicate for which he was an editor wasn't right for the story. Laura's Letter #4 of May 20 was postmarked 3:30 PM May 21 at New York's Times Square Station, and time-stamped 3 PM May 22 at Chicago's Main Post Office. Remarkably fast, a word-for-word account drawn directly from this letter was published—without identifying her—in the May 23* Tribune *under the head "FATE—Letter Tells How Shell Killed Nurses As They Stood In Group".*

Twenty-three years later, in the September 1940 issue of Esquire *magazine when a new World War was underway in Europe, Norman Klein published his own lengthy eyewitness account of the* Mongolia *accident. Klein, a reporter for the* Chicago Daily News *serving as an ordinary private in Base Hospital 12, wrote numerous accounts of the unit's adventures under his by-line and despite wartime censorship during their stay in France, examples of which we shall see later. The* Esquire *story was headlined "Two Heroines Unsung—Timely reminder to brass hats, that our first A.E.F. casualties in World War I were not killed by the enemy".*

Here's a dramatic excerpt from his story:

The stern gun blasted. After a deafened moment Miss Emma Matzen, of Columbus. Nebraska, one of the nurses on the boat deck, leaned over to Dr. Milton Mandel.

"Doctor," she said, "something's happened."

"Where?" he said.

"Those girls—aren't they hurt?"

He turned and saw Miss Wood. She was sliding awkwardly from her deck chair. Beside her, Mrs. Ayres already sprawled on the deck. Mandel rushed toward them. Others hurried up. They saw that Miss Wood had a frightful bruise in the head. Mrs. Ayres had a wound over the heart. Then a nurse, aghast, turned to Miss Matzen.

"Why, Emma." she cried, "you are bleeding too." Miss Matzen said weakly, "Am I?" They gave her treatment—she was the first of many thousands the unit succored in those two horrible years that followed. But they could not help the other two. The women were dead.

A girl, two years out of [nursing] school, began to cry.

"It's so awful," she said. "Now Helen Wood's the third to go in her family. She told me on the train how her two brothers enlisted from Scotland in 1914—in August, I think. They were killed in France. And now Helen —"

Yet this was not the first time these nurses had encountered horror. Many of them, including Laura, had been summoned from Cook County Hospital on the morning of July 24, 1915, after the overloaded and unbalanced Lake Michigan excursion ship Eastland *overturned in the Chicago River with a loss of 844 lives, many of them women and children.*

Still, the nurses' indomitable spirit, to be demonstrated often over the coming months, first showed in Laura's next diary entry, made only for herself. The two nurses who were killed were buried with military honors, accounts of which will be found in the Appendix.

Letter #5

Laura to Erle:
S.S. Mongolia *stationery*
Mon. 11 P.M. [May 21, 1917]

. . . I received your second nice letter this afternoon when Pauly came back from seeing Emma into the Naval Hospital. And I did like it so well. I needed it for it has been a very bad 30 hrs for us all. (Pardon the abbreviations and the scratching. They are gathering up the mail & I'm so afraid I won't get this in)

Ernest K's note was very funny & the English was worse. And he the editor of the paper.

The X Ray of Emma's wound showed nothing but of course there is infection & necrotic tissue and she is the grittiest woman you ever saw. They carried her over the side and she smiled and waved to us as long as she was in sight.

We've anchored in the middle of the Hudson but are not allowed shore leave—lest someone might desert, I suppose.

They are loading up new ammunition & putting on more coal & we'll probably put to sea again tomorrow. I hope so.

Tonight the privates were invited up & we had a sort of sociable to get acquainted better & also to get sort of cheered up for a fresh start.

I drew a very nice young man as my partner. He is a graduate of Northwestern and has been buyer or something or other at Butler (?) Bros. We had a nice stroll around the deck but I couldn't help wishing for some one else in his place and he probably was wishing likewise. And then I excused myself to come in & write this. . . .

Pvt. Baker's Memoir:

We arrived back in New York on Monday at 6:30 a.m. Many of the crew wanted to desert ship because of superstition and thought we were an ill-fated outfit. It did look that way at that. They said they also lost a black cat and that meant bad luck. To prevent any from deserting, we anchored way down the bay and only the dead, wounded and some mail were taken off. They then picked up a lot of our shells from the hold and took them to the proving ground to test them. We learned later that all were found OK. We thought that Hun trickery was at work in our munitions factories.

USN gun crew on Monrovia.

Lazing on Monrovia.

Naval gun crew on Monrovia.

Photo # NH 91268 S.S. Mongolia (1904). She was later USS Mongolia

The S.S. Mongolia.

Letter #6

Laura to Erle:
*[**S.S. Mongolia** stationery; no salutation]*
Tuesday Noon [May 22, 1917]

Now, my dear, don't think for a moment that you are going to receive such an endless string of epistles as you have in the past few days. Each one seems to be the last and then when I see another opportunity to send one ashore I immediately give in to a strong desire to talk to you again.

Today is gray and chilly and we aren't allowed shore leave. So we have walked and walked and sat and talked and that's about all. I have read nothing except the few newspapers that have drifted aboard on passing tugs and launches.

I hope you've been able to read my scratching even tho it is horribly illegible. Dr. Besley went ashore this morning with the last bit of mail and I have no idea whether this will go or not. But then if it doesn't I'll add to it occasionally & send it when we reach England. . . .

. . . we are anchored amidstream about a mile below Grant's Tomb. We can see that very distinctly as it is high on the cliff. The Hudson must be very beautiful in clear weather. The banks are very high and picturesque. But of course where we are is not beautiful unless we add greatly to our imaginations.

We have excellent food—in fact much more and more of a variety than we could expect. I'm agreeably surprised even tho I had always heard that the food on the boats is excellent. The boys told me yesterday that their food is not at all bad anymore and their quarters (third class) are fair and in fact everything is really better and they are allowed more privileges than they had expected. They were finally given their uniforms yesterday and they look so nice and so pleased with themselves.

There is quite a number of army doctors aboard who are not of our unit. They are to go right to the front and get some experience so as to be ready for our troops when they come. That doesn't sound as if the six units* would be the only ones sent but of course as every one is clamoring to go it will probably amount to the one coming that can put up the strongest "drag".

I don't think there's a person in the unit who would go back even if were (?) given the opportunity (which we aren't). We are only glad of the chance to go but terribly sorry to leave any of our number behind. And we all know that it's simply fortune of war that we are here and they are not.

I hope my letter to you after the accident wasn't too hysterical. I'm afraid it was but then you know I'm not really a very calm person and you will understand something of what it meant to us all.

It surely was thoughtful of you to send the clippings to Bertha. I know the folks will all appreciate them more than I can or they will tell you.

Last night I wrote them more fully when we found everything had been published anyway. I almost wish for the sake of the home folks that it had been more suppressed. But that also is out of our hands.

They took on more coal and provisions and changed the ammunition yesterday and last night so now we're just waiting.

There seems to be a very nice bunch aboard—the only swift looking people are the two stenographers. Everyone else is sort of responsible even tho foolish and apparently care free.

I understand that this is built as many of the Pacific boats are—not much room for passenger and lots for freight but that she is safe and steady and slow.

By the time we reach our hospital we'll all have forgotten how to work I'm sure. We are surely being lazy. We have some rather good musicians among both the nurses and the Privates. If the Doctors have such accomplishments they aren't airing them. They do not look musical, I'm sure, but one never can tell. . .

**The six U.S. Army Base Hospital units organized in 1916 from leading medical schools.*

Pvt. Baker's Memoir:

Wednesday, May 23

At 5:30 a.m. we again left New York. We had been given our Army uniforms and packs on the 21st and 22nd and all the boys were chasing around trading with each other to get a fit. When we stopped back in New York, I mailed my blue serge suit home in my laundry bag and it got home OK, I heard later. Many men just traded off a $50 suit for a hot baked chicken from the first cabin chef, or a couple of pies. The crew would wash clothes and take civilian clothes for pay.

Men were called to reveille at 6:30 a.m. and put through a drill and calisthenics. After breakfast the boys had boat drill, boxing and potato races, etc., to pass the time away.

The nurses like the boys. Most of us are from college. Our packs weigh sixty pounds, I believe. They feel as though they weigh a hundred when carrying them. We are to get $1.10 a day, regular Army pay.

Thursday, May 31

Our full pack consists of 1 Pup Tent, Red Cross Small Kit, Aluminum Combination Plate/Fry pan, Canteen, Condiment Can for hard tack, Tent Pole, Barracks Trunk—20" x 16" x 10", Kit Carrier, and large Bowie Knife with Sheath. Our personal equipment consists of 1 Hat, 2 Pairs Brown Shoes, 2 Shirts O.D. [*olive drab*], three Suits Underwear, 1 Pair O.D. Breeches, 1 Coat, 1 Pair Leggings, 1 Dozen pair laces, 6 Pairs socks—light weight, 1 Overcoat, 1 Hat Braid and 1 Blanket.

Every one is trying to roll his pack and it is some stunt. Some birds get their packs up too long and it scrapes their legs, others roll theirs like a short fat log, ten inches thick. We'll learn if we don't get torpedoed too soon.

Letter #7

Laura to Erle:
Thursday the 24th

Laura's next letter, written on Mongolia stationery over several days during the voyage to England, was the first of many to be censored over the next year and a half.

Such a long, quiet dreamy day as I've had. I walked and walked till I felt I'd had sufficient exercise and then I sat down or rather ended up in a steamer chair and read a silly book that required no thinking at all and listened to the water swishing against the sides of the ship. The water is very, very moody and I think I like it very much. This morning it was all steely gray and swishy and cold. This afternoon the waves are much larger and not as gray and look so entirely different.

The swishing sound is so familiar but I'm sure I've never heard it except in imagination when I've been reading some old sea yarns.

I've been wondering all day what you are doing and thinking. Twice I've gone below to read over the two letters I had from you in New York. It will be so long before I have another because it will take us till at least a week from next Sat. to reach England. When we come into the war zone we will zig-zag and that will take longer.

The guns are trained on every vessel sighted lest she be a raider. It's quite exciting—this running a blockade. I had never thot I'd be engaged in such a thing.

Yesterday morning at 5 o'clock we weighed anchor and were towed out to the entrance of the harbor and then dropped anchor again to wait for the fog to lift. We got under way at about ten. The sailors were close beside their guns because they had heard that [a] German raider was about disguised as a tramp steamer and also that a submarine had been sighted along the Maine coast. This latter information we gleaned from a sailor lad who loves to talk so I can't really call it authentic.

The chaplain is sitting nearby with the social committee planning a card party. He comes from the Englewood Methodist church. Do you think war has been a broadening influence? But he is all right. Everyone likes him.

Did I tell you that we have twenty-two sailors and Navy lieutenant? They are the "armed guard". It is volunteer work and they do [not] undress at all as they must be ready to man the guns at an instants notice. It's sort of weird, this slipping along in the darkness without a light.

Tonight we went out on deck and watched the Phosphorous in the waves. It was beautiful but I couldn't watch it long for some reason or other. The waves are running higher tonight and it's beastly cold.

We had a drill yesterday. We all put on our life preservers—mine must have been intended for a very fat man—and were assigned to the various boats. There are six women in our boat and about forty men. We should be well cared for. . . .

Letter #7, continued:

May 25th

Oh honey! The moonlight on the water!!! and I have wanted you so badly all day just to enjoy everything with me. The ocean has been blue and smiling all day and this evening the sun sank right down into the water and then the little moon began to shine and the stars came out and it's perfectly gorgeous!

The gorgeous I simply slang because I can't think of an adjective fitting the occasion. Do you s'pose we'll ever get to see such things together??

Mac says, "Isn't that beautiful?" and the other girls only laugh at me so you see —

Letter #7, continued:

Tuesday the 29th

. . . I've been dreaming and thinking and thinking and dreaming and watching the water. Today we passed two ships but they were only columns of smoke on the horizon. The ocean seems so very large—so much larger than I had thot. I had no idea we'd go for a week without seeing a ship closer than that.

We've had a few social doings—dances, card parties, races, etc. but as you know I'm not much socially and I'm sure I had a lower score than any one else at the card party. I much prefer to sit and read and entertain myself.

The navy lieutenant looks very much like Ferd Haworth only he is better looking and I'm sure is more intelligent. The sailors are interesting to me. One boy especially does love to talk and I'm sure you'd enjoy his slang as much as I do. He calls the British sailors "Limies" or "Blighties" and isn't at all fond of them.

We are going to have our pictures taken today. Many of the officers have their kodaks. If I find I can use one in France—you see we don't know yet how we'll find things—I'll get me one there. We are hoping we may have a day or so in London. But we are also anxious to get to work.

I haven't been sea sick at all and I feel very fortunate. None of our family were ill. And we haven't been inside at all. The only shower we had was while we were anchored in the river. But, my dear, you should see my face! It is the color of a lobster and now my nose is desquamating as if it had scarlet fever. But I'm happy and care not at all that I have the reddest hair and the reddest nose on the ship [***sunburn***]. . . .

But we have three good meals a day besides beef tea at 11 and tea with lemon at 4. . . .

Letter #7, continued:

Wednesday [May 30]

The nicest, drizzliest, waviest day. I've been out on deck all day just a looking. A British freighter passed us today and we were all so pleased to really know that we aren't alone on the Atlantic after all. Fifteen of the privates have been detailed to stay up with the sailors and carry ammunition. We enter the danger zone at 9 P.M.

Thursday Night

. . . Tonight was beautiful. Elizabeth Cleveland, who is very companionable, and I went up on the upper deck and wrapped up in steamer rugs and looked and looked and looked some more. And I tried to imagine what you were doing but I find it so hard to transpose time that I gave up and just looked some more again. We haven't had many clear nights—two I think—as there is most always fog if it isn't cloudy.

We've been in the danger zone some twenty six hours but have seen nothing so far except a tramp steamer away on the horizon. . . .

I haven't been sea sick at all and I feel very fortunate. None of our family were ill. And we haven't been inside at all. The only shower we had was while we were anchored in the river. But, my dear, you should see my face! It is the color of a lobster and now my nose is desquamating as if it had scarlet fever. But I'm happy and care not at all that I have the reddest hair and the reddest nose on the ship.

Submarine Duel

Pvt. Baker's memoir:

June 1

The big waves blew up on deck in sprays. A fellow walks along like he is drunk trying to buck the roll of the ship. Went up to the front of the ship again and wind and waves were great. Wind would nearly blow one off his feet. Standing on deck and looking as far as one can see, the ocean looks like a great group of hills, dark blue in color and smooth valleys in between—a wonderful sight. We sat on deck with nurses last night. They are not afraid, but anxious.

At noon, 12:00 o'clock, as we were down in our cattle pen eating, we heard the mournful blast of our whistle, three times which means "Sub" attack. The boys beat it for the ladder and scrambled up. A lot of us kept eating, thinking it only a practice drill again. We could not all go up a ten-foot ladder at once anyway. However, when we heard the excitement on deck a second later, we also became a little anxious to get up. The hands of a lot of boys were stepped on.

When we got on deck we found a heap of excitement and everyone had a life preserver on and was standing in line by his boat. Bang! Bang! go two of our guns. I was in my stateroom getting overcoat, preserver, canteen and filling my pockets with oranges and chocolate so if we were caught in an open boat, I'd have some provisions with me anyway. Each boat had hardtack and some water in it.

I was eleventh in line for boat #14 and we were facing the left rear gun which was banging away. Every shot would make one's head rattle. I held my ears when they fired. Everyone was ready to be lowered in a boat if our ship were to hit and sink. Our big 6" aft gun fired about five shots at something which we were to learn a little later was the periscope of a Hun sub. It barely came up and in a half minute it had lowered again but one of our shells hit it square and exploded about one hundred fifty yards off the rear of our ship. Some of our boys helped the sailors hoist up the 105 lb. shells to the guns, giving them a lot of new experience.

One torpedo missed the rear end of our ship by fifty yards.

The Captain says that three torpedoes were fired from three different Subs. One missed our front end and one fell short. We, of course, were zigzagging all the while to throw off their directions.

Captain Rice sent out wireless distress calls and just as firing ceased, we saw the most wonderful sight we ever hoped to behold—a British destroyer, off our stern ten miles, and racing to our assistance. We felt like the hand of God truly had answered our calls. The destroyer came tearing through the water at full speed and was soon making circles around us. It went back to the spot where we had hit the periscope and reported a lot of oil on the water so we later got credit in London for sinking a submarine.

The Allies did not commence the practice of ship convoys across the Atlantic until September 1917; so single ship crossings like the Mongolia were especially vulnerable to German submarine attack in the early months of America's involvement.

Letter #7, continued:

Friday PM [June 1]

Friday Noon we really had some excitement and your little friend was almost submarined—only she wasn't. You see it was like this. We had been having a shower and I was on the upper deck under the bridge watching the clouds and the curtain of rain come and go when I heard the man on the bridge shout, "Port side abeam," and saw sailors rush to the guns and begin to aim. I started over to see a real live submarine but couldn't see a thing when the danger whistle blew and I suddenly realized that I'd best be getting my life belt according to orders. I ran down the gangway then and to my stateroom, gathered up my sweater, blanket & life belt and came back to the lounge. Every one was quiet and while some were excited I think very few were really frightened. We fired five shots and then some one said "There's our convoy ["] and sure enough away on the horizon came the British torpedoe boat destroyer. The only torpedoe fired at us went about fifty feet astern. But maybe everyone didn't yell when that British boat came up. And some one below played "Star Spangled Banner" and then "God Save the King." It sounds like a story, doesn't it, but it's really so. And I feel as if I'd lost the chance of a lifetime by not seeing it and I certainly should have when I was where I was.

The first mate said he saw the periscope for just a minute and she was so close that it couldn't have [been] more than 400 ft away.

And you should see our convoy zigzagging across our bow like a gay little puppy, only much more graceful.

We land early tomorrow morning. I hope this all goes through without being much censored but if it doesn't no harm is done.

The water is green this afternoon instead of blue or gray but we have not yet sighted land. I don't know when I'll have time to write again but anyway this is almost a serial isn't it?. . .

Morning [Saturday, June 2, 1917]

We can see the green fields of Cornwall with the dividing hedges. Merrie England!

Nurses on Monrovia.

While the SS Mongolia was still at sea, the following memorial verse was published in Base Talk, BH 12's one page on-board newspaper:

IN MEMORIAM
TO HELEN BURNETT WOOD END EDITH AYERS

Two blurs of cloudy white,
The whirr of wings,
Two dim forms, disappearing
Into the Infinite blue.
Why is it that as souls pass
There is no sign?
No sign, nor whisper nor caress
To those who weep bereaved?

Mayhap were one so faith full
As to live each day the thought Divine
The Soul's swift winging flight through space
Might be a sign from Heaven
And visible.

Can it be there is no greeting
In after times,
No smile, nor kiss. nor hand-clasp
For those who stay behind?

Ah, yes, we all have faith in this—-
That when the time shall come
And the summons bids us follow,

That 'mid the everlasting stars of heaven
Our loved ones wait with outstretched hands
To give us welcoming!

And that, our little journey finished,
Our earthly task well done,
We'll hear the sound of voices—eager, happy.
Welcoming us Home!

— ELIZABETH CURTISS NOLAN

England and Wartime Sightseeing

Diary of Dr. M. Pinson Neal, BH pathologist:

June 2—Put into Falmouth Harbor at 7 a.m. Left ship at 3 p.m. Put up at Falmouth Hotel.

June 3—Left Falmouth over Great Western Ry for London at 11:30 a.m. Arrived London 8 p.m. and to Hotel Curzon—Philadelphia and St. Louis units yet at Curzon Hotel told us that the papers had published a report that our boat had [been attacked] and all aboard had gone down in the mid-Atlantic due to being torpedoed.

Letter #8

Laura to Erle:
London
6-4-17

Dearest Buddy:

I've just learned today that our ship was reported sunk. I do hope there was no such report at home because you see we all [are] perfectly all right even tho the ships ahead and behind us were all torpedoed. We feel very fortunate indeed. But, honey, it makes me feel so scared when I think of you coming. I know it's absurd and I wasn't the least bit frightened for myself but some way it's different when some one else is doing it.

We landed in Falmouth and found it such a delightful spot. We were allowed to see the castle (Pendennis) and everyone was charming to us. I know you would enjoy the English gardens and hedges and the queer stone houses and the tiny little fields. We were all so perfectly delighted with everything. The next day we came to London on a special train. That is the officers and nurses did. The privates were sent direct to Blackpool wherever that may be. We did so hate to have them go but of course we expect to meet them in France.

Today we've been to Madame Bussard's [***Tussaud's***] to see the wax figures and to the Tower and have had a drive around town. I'll not begin to tell you all I've seen because, you know, I couldn't possibly do them all justice and you will probably see them some day yourself. The Tower is surely very interesting.

But you should see me get all balled up on the English money. I'm so stupid about that and the streets, too. I'm always lost and never know how much money to pay or how much I have. I do not like London very well but perhaps after while I shall.

We are not sure just when we'll go to our final destination but we hope we won't be here too long.

Thursday we are invited to the American Embassy & of course we are going.

Almost all the men we see are in uniform or else are too old, or too young or are physically unfit for service. We [see] so many crippled soldiers that we are almost hating to get into the hospitals.

Every one seems to enjoy our talk so much. We don't need to tell them that we [are] strangers or that we are Americans. They all know that. . . .

Laura's passport photo, taken in London, May 1917.

Letter from Erle to Laura's sister, Bertha Huckleberry:

Editorial Rooms

Western Newspaper Union

Newspaper Service

210 So. Desplaines St.

Chicago, June 6, 1917.

Dear Miss Huckleberry, —

It begins to look like that little sister of yours is quaffing the real wine of life right regularly now. Witness the enclosed clipping from this morning's Chicago Tribune.

I would give my Sunday shoes to have been with the folks aboard the Mongolia during the encounter with the submarine. Such luck doesn't come my way, however; when unit No. 11 goes to Europe the voyage will be about as exciting—well, let me see: as exciting as doing a week's wash and ironing the same.

When our unit will go remains a mystery, I suspect, even to the people at Washington. We may be sentenced to stay in this country and nurse footsore rookies at the training camps next fall and winter. Wouldn't that be a heart-breaking anti-climax? It would probably drive every unit officer, nurse and private out into the garden for a feast of worms.

Sincerely,

JED

MONGOLIA AND SUBMARINE IN DUEL; NEITHER DAMAGED

LONDON, June 5.—The American steamship Mongolia fired four shots on June 1 at a German submarine which discharged a torpedo at the liner. Neither the Mongolia nor the submarine was damaged. The Mongolia was 250 miles from the British coast when the submarine appeared and fired a torpedo. By skillful maneuvering the steamer escaped it and opened on the submarine with her guns. Four shots were fired. The submarine fled, although unhit. The Mongolia proceeded on her voyage and has arrived at a British port.

It was the Mongolia which fired the first gun for America in the war and sank a German submarine. After leaving on the voyage just completed the liner was put back to port on account of an accident to a gun.

• • •

On a voyage to England in April the Mongolia fired on a submarine in British waters. On his arrival in London Capt. Rice said there was no doubt the submarine had sunk. It was reported unofficially later that the captain of the U-boat had been killed and that the craft had been damaged badly, but had reached Kiel. When off Nantucket Island on her latest voyage two nurses were killed and one was injured by a defective shell fired during gun practice.

Letter #9

Laura to Erle:

HOTEL YORK
Telegraphic Address "ESMERO.OX.LONDON"
Telephone Nos. 4100-1-2-3 MUSEUM
BERNERS STREET
LONDON W.1.

June 6, 1917

Dear Buddy:

I almost made the beginning very sweet so as to frighten the censor away and then I decided that as he'd probably wade right thru anyway I wouldn't waste time on confectionary.

We are fairly busy these days seeing Londontown. We don't know yet just how long we'll be here nor just exactly what to see so we aren't making as good time as we might if we had planned this jaunt.

Yesterday PM we went to Westminster Abbey and heard a service. I hope you come this way and have the opportunity to hear the organ and the voices in those dim high arches. It was beautiful. And besides being beautiful it is very interesting, you know.

Today we went to St. Paul's Cathedral. It was beautiful also but is entirely different from the Abbey and not quite as interesting. Did I tell you about the Tower? It is so full of memories of famous prisoners that I wish I had reviewed my English history a bit before I came.

Have you received the two other letters I've sent? One was mailed from the boat before we came ashore. Please don't forget to let me know. You see I haven't learned yet what will be censored and what will pass and so haven't been the most discreet in my writing.

And did you hear in Chicago that our boat was sunk? We heard here that it was published in the London & N.Y. papers and I've been so worried about the folks hearing such things. I can't think how such a report could be started when there are enough bad things happening all the time without imagining some extras. . . .

We see so many, many crippled and mutilated soldiers on the streets. And almost every young or youngish man is in uniform and the ones who aren't we wonder why they aren't.

You'd like Falmouth I'm sure. It is so quaint and charming. We were all much happier there than here but perhaps it was because we had just landed. But everyone is lovely to us and we're seeing many things.

We want to go to work now. I've been loafing or semi loafing since May 1st and I'm afraid I'll forget how to work.

You should see the shoes we have to get. They rival your army shoes I'm sure. The soles are almost an inch thick and they are terribly heavy.

A man who writes for Scribners—can't think of his name at all now—was here to call on Miss Urch to gather material for a write up about the American units. I don't know whether he got any or not.

Tomorrow we go to the American Ambassadors to take tea with Mrs. Page. We have had several invitations from different individuals or societies but have accepted very few as our time is limited and then you know we didn't exactly come over just to attend afternoon teas.

I've so much to tell you about everything. In about two years from now I'll probably talk your right arm right off at the shoulder and then start in on your left. But perhaps you'll cripple mine likewise with your experiences. . . .

From an unknown newspaper

U. S. Doctors and Nurses
Lavishly Entertained

London, June 7—The American doctors and nurses comprising the units from Chicago, Philadelphia and St. Louis are being entertained here on a lavish scale this week. Nearly every hour of their time is being demanded by their English, Canadian and American hosts, who seem to have arranged in their honor no end of receptions, dinners, teas, theater parties and sight-seeing trips. . . .

Group photo of BH 12. Personnel at Lipton's estate, with his signature

<u>*Letter #10*</u>

Laura to Erle:
Sealed with **Opened by Censor** *label; hereafter, almost all Laura's letters, even after the war ended, were marked as "censored."*
London, England
June 10, 1917

My dear Friend:

We are still busy sightseeing and being entertained. I don't remember what I've written before so if I repeat I hope it won't bore you too much. We've seen many of the important sights of London and many smaller interesting ones.

One of the latter was "Ye Cheshire Cheese," the old inn where Ben Johnson and his friends used to gather. It was on a funny little court off of Fleet St. and was in the rear of a saloon so I did not get to eat or have tea there as I should have liked. If I ever come back here I'm going again with a less particular crowd.

And did I tell you about going down to Stoke Park, Ipswich, to visit Lady Henneker Heaton, whoever she may be? Frances became acquainted with her in the lobby of the hotel and she invited her to come down and bring ten friends with her. I was one of the ten. We had a two hours ride on the train (two compartments reserved for us by the president of the road) and were met by carriages and driven out to the place. It reminded me of the estates pictured in the English magazines. The outside of the house was not prepossessing but the inside was delightful. There were such great light rooms and so comfortably furnished. The bedrooms (there were thirty in all) were huge and most of the beds four posters with canopies. The library was inviting with its Turkish red upholstering and its many books and curios.

And of course the family portraits were not forgotten. I know you would have enjoyed seeing it all.

Then today Sir Thos. Lipton entertained all the doctors and nurses. He a delightful old fellow and he certainly can entertain. One of the girls said she felt as if she had been to a Christmas tree and it was something like that. We were given red, white & blue bouquets, American flags, fed on lemonade and ice cream and then turned loose in the grounds or rather park for a while. We had moving pictures (please watch for us) speeches, songs, readings and tea with strawberries and such absolutely [***word missing***] cream, cakes, bread & butter and sandwiches and were finally sent home with a box of candy for each lady and cigarettes for the men.

We are beginning to get anxious to get to work and perhaps it won't be much longer. We have not yet received a definite address so I'll let you know as soon as I find out. I'm not sure whether we'll come under British or American Expeditionary Forces but of course I know that I can't give the location anyway.

I keep wondering how many of these letters get thru and what is censored. We are all beginning to long for news from home. I even went down to the Cook Agency and asked for mail. They laughed at me of course but I thot it wouldn't hurt anything to ask.

Anyway Capt. Porter told us that three mails have come in since we arrived and not a one of us has had a letter. I'm going to reform and write to Miss Stodola and every one I know so that I'll get some letters. I've certainly written more in the past three or four weeks than ever before in the same length of time.

I have many small things to do tomorrow A.M. so I must stop this monologue.

Goodnight.

Laura H.

From an unidentified Chicago newspaper

Chicago Red Cross
Is Entertained by
Sir Thomas Lipton

LONDON, June 10.—Sir Thomas Lipton entertained at his country place this afternoon the Chicago Red Cross unit of thirty surgeons and sixty-five nurses. Among the other guests were Prof. Richard P. Strong of Harvard and the Italian bacteriologist, Dr. Castaleeani, both of whom were associated with Sir Thomas in combating the typhus plague in Servia.

One of the features was an entertainment by theatrical artists, who sang American patriotic songs. Refreshments were served in a tent and there was speechmaking.

Sir Thomas said that Prof. Strong and his coworkers had cleaned up Servia after an epidemic which cost 150,000 lives and delayed military operations for six months. He paid high tribute to the late Dr. James F. Donnelly of New York, who died while fighting typhus.

Dr. Strong spoke highly of the work inaugurated by Sir Thomas in Servia and expressed appreciation for the aid given to the American Red Cross there.

Maj. Collins spoke for the unit and Maj. Besley in behalf of Chicago. The party's return to their hotel in a long procession of motors, waving American flags, attracted much attention along the way.

V. EAGERNESS TO DRUDGERY

WELCOME TO DANNES-CAMIER, PAS-DE-CALAIS, FRANCE—AND CENSORSHIP

SUMMER 1917: U. S. BASE HOSPITAL 12 REPLACES BRITISH GENERAL HOSPITAL 18

Memoir of Reserve Nurse Elizabeth Lyon:

After a week at the York Hotel, London, about noon June 11th, we boarded the train at Charing Cross Station, went to Folkstone, crossed over to Boulogne and traveled by lorry to 18th General Hospital, B. E. F. [*British Expeditionary Force*}, Camiers, arriving there about 10:30 p.m. The English sisters [*the British term for graduate nurses*] provided beds (i.e., boards on two-by-fours with army blankets spread over them) and a supper of what the Tommies called "bully-beef," bread and tea. They had not expected us that night. An entertaining side issue illustrative of the splendid spirit with which the nurses met all hardships occurred that evening. During the trip one of the nurses said to a friend, "Oh, if I only had the soft side of a board to lie on." Now as they viewed their beds her friend said, "There is your soft side of a board. Take to it." Which she did without delay.

Before retiring that night the English Matron Miss Williams and Miss Urch arranged the assignments and the nurses went on duty the next morning and the next day the English Sisters were relieved. We were only sixty nurses and there were about fourteen hundred patients in the hospital so the English left twenty-five V.A.D's (Nurses Aides) to stay until reinforcements could be sent over. In a few days we were "carrying on" as tho we had always been there to quote our English commanding officer, a proof of the wisdom of selecting groups who were accustomed to working together and convincing evidences of the adaptability of the nurses.

Diary of Dr. Neal:

June 11, 1917—Left London at Charing Cross Station over South Eastern and Chatham Ry—went aboard S. S. Onwart to France—all unit again united on boat—several other boats in fleet convoyed across the Channel by destroyers and Zeppelins. Arrived Boulogne, France, at about 7:30 p.m. Then taken by automobile to Dannes-Camiers hospital and home, General Hospital No. 18 BEF, a small city of tents. Heard guns fire all night. Some British officers still there to help us [with] our orientation and indoctrination.

Pvt. Baker's Memoir:

Monday, June 11

Our ship was ready to steam for France so we left Folkestone at 4:45 p.m. We sailed on the troop transport "Onward" with four other transports full of troops. We were accompanied by four destroyers, a submarine, two French dirigibles and a couple airplanes [*Note that ship convoys of merchant ships and troop transports were herded by protective warships across the English Channel, even though at that time the practice was not employed for Atlantic crossings*]. We had little fear of subs with all this protection. The Channel was very quiet all the way. It only took us from 4:45 to 6:00 p.m. to make the 25 miles across the Channel. We pulled into Boulogne and all one could see along their waterfront were thousands of sailing vessels—fishing smacks. We soon learned that this forest of masts was the real means of livelihood for the people of Boulogne, France, a great fishing center.

We unloaded without any welcome (the people did not know we were coming) of any kind, we marched up through the town followed by the women and children trying to sell us oranges, etc. The town is on one big hill. After three or four miles uphill we came to a large British convalescent camp. We were welcomed by a British Major and a lot of Tommies and Scots. We here met face to face, the first great bunch of wounded men, over 2,000 of them. Many were wounded at Messines Ridge where a major drive was on after the English had tunneled clear under the high ridges fortified by the Germans. After putting tons and tons of dynamite under the German positions, the English blew it up and the poor Germans were blown to all parts of France and Belgium. Then the English and New Zealanders charged the ridge, following their heavy cannon shelling, and took the ridge. There was terrible fighting during this advance and these men here had just come in from that scrap. We spent the night here in tents and had a good rest. We could hear the heavy guns at the front very clearly and we now began to realize there was a war on and not so very far away.

Diary of Dr. Neal:

June 12—Dannes-Camier—city of tents—five General Hospitals—10,000 beds—some 2000 assigned to each unit. General Hospital No. 20—Boston (Harvard unit in British Army); General Hospital 4; General Hospital 11—Boston; General Hospital 24—and ours.

June 13—One other hospital, Portuguese and a British Officers Hospital. For initiation, convoy in at 10 p.m. The 13th Machine Gun training area very close to our unit. Constant rat-tat-tat-tat all days.

General Pershing landed in France—arriving in Paris late a.m.

Batman services for officers—pain (?) Charmers! Smith! Tea with bread and marmalade 4 to 6 p.m.—so-called "social hour"—a daily event—when all non-emergency work ceased. Went to British Supply Station at Is-Sur-Tille for equipment-supplies.

June 15—Took over charge of Officer's Mess from British officers. Former British officer in charge of laboratory, General Hospital 18, left in p.m., leaving me with a little equipment for clinical, pathological and bacteriological but nothing for doing tissue pathology but a microscope.

Letter #11

Laura to Erle:
"Somewhere in France"
June 12, 1917

Dear Buddy:

At last we are settled. We've been twenty six days on the way so we're already to go to work. We arrived last night at nine thirty and have found things very comfortable—in fact much more civilized than I had thot they would be.

I have been assigned a fifty bed ward that is not at all full. But of course it may be filled over night as they come and go. I have two orderlies. Their ways of doing things are somewhat different from ours but of course we will learn in time.

We are called "sisters" and only the V-a-d's (Voluntary Aids) are called nurses. We are finding much difference in our customs, also.

Today I received your "fat" letter of May 22nd. I was so glad to get it. I've read it about six times. It was the first since we left New York.* When some of the others received mail in London I was heart broken that I got nothing. And I haven't heard from home at all since I left Chicago.

There is so much I'd love to tell you but I cannot. Miss Urch is going to censor our mail and you know what a cross lady she is. She told us today that some things that shouldn't have reached the newspapers so we are to tell our correspondents not to publish our letters. This is a warning not to sell any of my stuff to either the Tribune nor the Century Magazine. I'd hate terribly to see any in the Sat. Eve. Post. Beware!!!**

...

We were better pleased with England (except the mutton and fish) than the bit of France we've seen. But then, you see, we've seen practically nothing of this country. We can't judge fairly yet. . . .

** The three-week mail transit time would turn out to be typical for U.S.-to-France letters for Americans in 1917-19, although some took a week less and others a week or more longer. During World War II, single-page "v-mail" letters (photographically reduced in size for lighter and smaller transportation loads) saved valuable space on ships, and relatively swift airmail became the norm during the Korean and Vietnam conflicts.*

*** This "warning" no doubt was tongue-in-cheek. Laura knew full well her sweetheart was a "newspaperman" (as they were called before "journalist" came into vogue) and would protect his sources. When Laura reached the site of Chicago's 1915 Eastland disaster, mentioned above, she telephoned Erle at his newspaper office to tip him off to the news.*

The letter's quality stationery—unlike the wartime military issue used later by Laura—was embossed "Truslove & Hanson Ltd. 153, Oxford Street, W" with envelope stamped in red, "PASSED BY FIELD CENSOR" with image of British crown, "On Active Service" written in right corner, initials "L.G.H." in left corner, "OPENED BY CENSOR" seal around top. This was Laura's first letter to Erle from the hospital in France, and carefully avoided, under censorship, telling the location.

The New World Meets Old Europe

Pvt. Baker's memoir:

June 12

We were tired out last night and needed a lot of rest. At early dawn, as we were all sound asleep, we were awakened by an awful noise. We jumped up and looked out from under the tent to see from whence came all the devilment. We were surprised to see a lone Scot blowing his bagpipe for all he was worth and parading all around camp. There was no need for this stunt as we had first call at 6:15 or two hours later. If he were one of our men, he'd have died in his tracks but I guess the Scotch just can't subdue their surplus spirits and take it out on a bagpipe and us poor goofs. Well, after laughing and cursing him, we tried to go back to sleep while this guy kept on playing. This must be an ancient custom for these warriors.

After spending the day in camp, we marched down to the train and at 5:00 p.m. we left for Camiers, 14 miles south of Boulogne.

About 6:15 p.m. we arrived at our own hospitals with about 2,000 beds. We were assigned to large tents, or marquees, for quarters and next morning we were all lined up to be assigned to our special duties. This sure was a strange performance. The Colonel would ask each man to tell what he was particularly qualified to do. About half the men had never done a day's work so it was not hard to tell the Colonel their special line. If a man said he was a good carpenter, he was made a cook, or if he said he was a salesman, he was put on the pioneer squad which was called the "Swiss Navy" for a joke. This was where I was assigned. Our job was to keep the camp in shape as well as carry in wounded when convoys came in. Most of the wounded here are British, Australian and Indian.

There is a unit from Boston across the street who came a week before us. We hospital volunteers—three units in all—are the first 500 Americans in France after America declared war April 6.

General Pershing landed at Boulogne on June 13 and came by our camp and saw our flag.

We are served British Army rations which is pretty fair grub compared to what we expected since England has been in the war for two and a half years. The menu consists generally of bread, tea, stewed beef or cold corned beef, jam or marmalade. At first we had cold, fat bacon served us for breakfast but we Americans immediately demanded that our bacon be fried, and it was after that. We also demanded coffee once a day and got that, too. The tea has practically no sugar in it. Goat's milk is used mostly for tea and coffee.

About 6:15 p.m. we arrived at our own hospitals with about 2,000 beds. We were assigned to large tents, or marquees, for quarters and next morning we were all lined up to be assigned to our special duties. This sure was a strange performance. The Colonel would ask each man to tell what he was particularly qualified to do. About half the men had never done a day's work so it was not hard to tell the Colonel their special line. If a man said he was a good carpenter, he was made a cook, or if he said he was a salesman, he was put on the pioneer squad which was called the "Swiss Navy" for a joke.

Letter #12

Laura to Erle:
France
June 16, 1917

Dearest Buddy:

. . . It is ten o'clock and it is still quite light. We've been up on the hill picking flowers for our tea tomorrow. We are supposed to invite the officers to tea on Sunday PM. Tea is a great sport over here. We drink it whenever there is the slightest opportunity. But coffee is not at all desirable as it is served here. . . .

Our hospital is principally under canvas. There are perhaps three or four wards in huts as they are called here. At home we would call them shacks.

The nurses or "sisters" as we are called here live in shacks but I think all the men are in tents. However I don't really know about that. We are quite comfortable. I'm not so sure how winter will be. Everything is much better than I had expected.

The Tommies are nice patients and I think I'll be quite fond of them. Some of them call our boys "Sammies."*

I haven't ventured to the village yet as I'm a little afraid because of my dearth of a French vocabulary.

The villages that we came thru are picturesque but not exactly sanitary. I think I prefer the English villages as far as cleanliness goes. . . .

** Americans hated being called "Sammies" (after Uncle Sam) and Americans from the South, especially since it was only 50 years after the Civil War, did not like being called "Yanks" because "Yankees" were Northerners. Probably the Canadians would not have agreed that "Americans" referred only to those south of their border. Yet "Tommies" seemed acceptable in both World Wars as a slang term for British soldiers. By the time of the Second World War, "Yanks" became a well-accepted moniker for Americans mixing with Britons and Empire forces.*

Diary of Dr. Neal:

June 18—So-called social hour for tea a big nuisance, nothing but an excuse to play "old lady"—have tea—and principally rest (coffee never right, etc.). None of it for me, too much of an American to become slave to another man's lazy custom.

June 19—Convoy* in at 2 a.m. Present system: breakfast 8 to 9, go to work at 9:30, stop at 12:30, lunch at 1 to 2, work until 4—tea hour until 5, dinner at 8. Doesn't go well with a man who came here to work and wants to do so.

** "Convoy" was what the British Army called incoming patient arrivals, via train and ambulance, of wounded and sick from the Front, or battlefield. In the Navy context Convoy meant a group of merchant ships protected by armed escort ships or aircraft, as when BH 12 crossed the English Channel from England on June 11.*

Letter #13

Laura to Erle:
France 6-20-17

My Dear Mr. Davis:

I know you will enjoy seeing the funny French villages when you come. They look as if they had come off a comic opera stage. The streets are mere alleys and the houses are so very different from our American ones. But the odor is not always pleasant and the streets and door yards are very dirty. They contrast with the spic and span English cottages and gardens.

Today it rained and rained and we were soaked before we reached home. You see this was my half day and I went on the bus (an auto truck with boxes for seats) to the next town [***Etaples***] and then took the tram to another slightly larger and cleaner town [***Paris-Plage***].

We had a fine time just prowling around and eating and looking. We ate twice so we'd be sure to have enough. The strawberries are delicious—much better than ours, I think. . . .

I'll never learn to speak any French at all here because there is always some one to speak English or to interpret. And I find the French money ***[based on decimal system]*** much easier to count than the English coins. . . .

Pvt. Baker's memoir:

June 20

At 6:00 p.m. we were all called out to help carry wounded from ambulances. It was heartrending and pitiful sight watching the poor fellows get out of ambulances maimed and wounded in all ways—many with trench feet. They were from the left side of the line near Messines. I sat until 9:30 last night talking with convalescent soldiers about their experiences and it sure looks like the Huns are going to fight on a long time yet. The English soldiers all say the Germans have plenty of men and guns and it looks bad.

The road to Base Camp 12. Beyond the hills in the background was the sea.

From unidentified newspaper

UNIT FROM CHICAGO
IN A WAR HOSPITAL
Body Organized by Dr. Besley
Conducting Battle Front
Post of British

By the Associated Press

British Headquarters in France, June 21—Quietly and with very little ceremony, so as to escape all but official attention, American physicians, nurses and enlisted men have taken over within the last fortnight six of the British great general field hospitals, releasing the English staffs for duty near their front. American surgeons have been rendering valuable aid also in the forward casualty clearing stations in the battle of Messines. This has been the most quickly realized aid the United States has given …

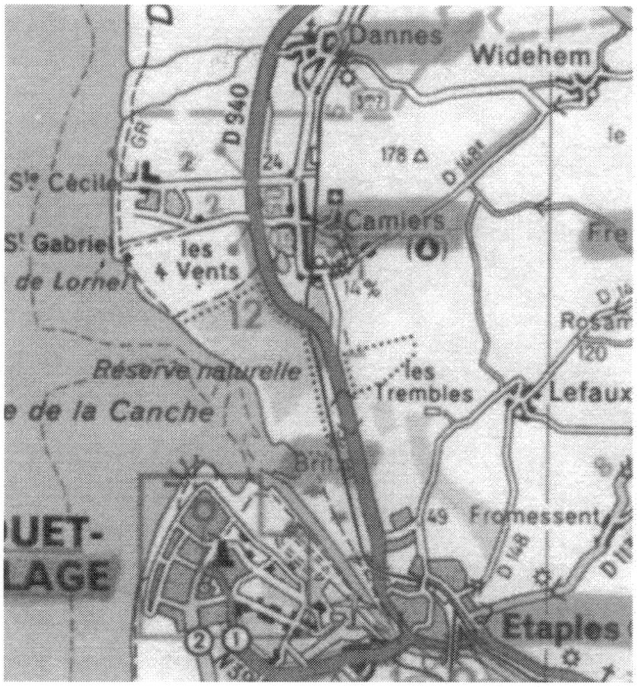

Diary of Dr. Neal:

June 25—Convoy in at 6 a.m. Boat sunk just off our coast (Channel) late p.m., hit by torpedo from a sub.

June 26—Planes over our area at noon, then put to sea. Convoy 2 a.m.– convoy 6 p.m.—convoy 11 p.m. Heard guns firing around Ypres battle line—a sensation of a far distant thud, an occasional more distinct report from one of the larger guns.

Letter #14

Laura to Erle:
France
June 25, 1917

Dearest Buddy,

. . . I do wish you could see our nice camp hospital. It is nicely arranged and we have flower beds and tiny vegetable gardens and rose bushes and everything. I think I told you before that the hospital is mostly under canvas—all except three wards and the Operating Theater.

Our quarters are in shacks but here they are known as huts.

Our patients come in in convoys. They go to the reception tent and then are sent on to the respective wards.

Tonight we got six new ones in our ward. They were not terribly serious cases and they were all so pleased to get into beds. I like the Tommies very much. They are so good natured and so gritty.

We have had just one since I've been here who wasn't as gritty as could be. And the first question they ask is, "Will I get to go to Blighty?", meaning England of course. You see if they will convalesce shortly they are sent from here to a convalescent camp & eventually back to the line but if the wound is fairly severe they are kept here but a few days and then sent on to England to better hospitals.

Tonight there is a gentle rain. Elizabeth Cleveland and I walked down the road about a half mile and back and would have gone farther except for letters to write.

I wish that you might see the sunsets and clouds and hills. They are all very obvious in the absence of high buildings.

I wonder if you are still in Chicago. In your last letter (May 25th) you thought your unit would be ordered out shortly. And I wonder where you were ordered to. . . .

Diary of Dr. Neal:

June 28—Convoy at 10 p.m. Many trains going through day and night—going toward Calais—all loaded with French soldiers and guns. Men say they are going as reinforcements on the Belgian border.

June 29—Up with Lab work 1st time. Many troops and gun loaded trains passing toward Calais. 1st U.S. combat troops land in France.

Letter #15

Laura to Erle:
France
July 1, 1917 [Sunday]

Dear Buddy:

. . . We landed here on the 11th and went to work the next day so we are getting down to business very well altho some things are unsatisfactory.

I'm sure I've told you at least six times that almost all the hospital is under canvas and that it will accommodate about 2000 patients altho it hasn't been full since we've been here. We must have room to put the new ones. They come in bunches called convoys. They have first aid on the field and then are taken to the clearing station where they are dressed or operated if need it and are then sent on to us or to other base hospitals in convoys. We keep them only a few days usually and then send them on to "Blighty" or else they go to convalescent camp and then back to the front.

We never know how many patients we may get from a convoy. Sometimes we get one and sometimes a dozen or more. They are lifted from the stretchers to beds, undressed and bathed and their wounds dressed if necessary. They usually sleep for about twenty four hours before they are rested. And then of course they are so glad to get into beds as most of them haven't seen one since they left home.

We have about a dozen Canadians in our ward now and I really feel as if they are our own people and they seem glad to see us, too.

I'm also getting used to seeing these terrible wounds. At first it seemed too dreadful. I do so hope it won't last much longer. I couldn't realize how horrible it was till we reached London and there the effects of war were apparent. . . .

I wonder where you are and if your unit has yet been called. In your last letter of May 25th you thot it would be ordered out shortly and that is over five weeks ago.

Doesn't it seem horribly long between letters? Already I've been away from Chicago almost seven weeks and that is almost one-seventh of a year. How do you like my Arithmetic?

Later

I had to stop and go to church just then. But now I'll finish this epistle before retiring. The church tent is about ten feet from our room so felt that I might as well be decent and go even tho' I could hear practically everything from where I am sitting.

But I really enjoy the services here very much. They are very informal but with a cordial spirit.

Night before last I went up to one of the concerts, the first I had attended. This one was held back in the hills. A little stage about 4 x 10 with a background of the flags of the Allies and the wild flower decorations faced a tiny valley. We sat on the hillsides and looked and listened. The music was nothing special but I did like to watch the gathering.

There were officers and "sisters" (the graduate nurses) and V.A.D.'s (corresponding to our Nurses Aids) and a conglomeration of Tommies, Australians, New Zealanders, Canadians, Americans and everything that could possible be a patient or connected in any way with a hospital.

There are concerts in the Y.M.C.A. almost very afternoon or night and some of them are supposed to be very good.

We pay about a dollar each into the mess every week and as a result have very good food in plentiful amounts. The ration that is allowed us by the British Government is sufficient to keep one from starving but you know we Americans are fond of eating and then we like different things than the English do. For instance they allow two ounces of tea per day and no coffee. So we bring coffee and are storing up tea even tho the English sisters had been buying extra tea every week.

A much larger allowance of cheese is given than Americans use and also of marmalade. But we are short on milk and butter and green vegetables, etc.

We brot Miss Lindsley, who has been matron at the Illinois Training School home for the past five years, as dietitian and she surely manages well. We probably are better cared for than any of the others unless they have some one her equal and that isn't at all probable. We pay only about a third of what the English women were paying and have far better food—at least we like it better.

A year ago tonight I returned from my vacation and nearly baked or rather stewed on the train. And here I sit with my cape on in the house and yet I'm not comfortably warm. . . .

YMCA building.

Pvt. Baker's memoir:

Saturday, July 1

I was transferred from the Pioneers, or "Swiss Navy" to Ward "O" which is full of gassed and sick men. I have as my pal, Orderly Will Peckham. We are under Captain Nadler. Miss H. Sparks and Miss Rettke, both of Chicago and in our Unit, are the nurses.

There is tent inspection by our Captain every morning and sanitary inspection every two weeks. We have an old brush with ten hairs on it and a long handle with which we are expected to do our sweeping. We cannot get our floor to look as though it has ever been swept so we get bawled out for it all the time.

We go to bed these nights warm and wake up about frozen. We pile on four blankets and an overcoat some mornings to keep warm.

I have all kinds of work in the hospital. I have to shave many fellows, change dressings, put on and take off splints, write letters home for the men, etc., etc.

Patients are coming in and going out every evening. Received some wounded after the big Messines explosion and from the second great fight at Ypres the latter part of May, June and first half of July. They tell of the awful hell going on at the line. There are British, Canadians, Australians, New Zealanders, Jamaicans, Scotch, Irish and Fiji Islanders in my ward.

When Mr. Lyons, the big Scot, and Mr. Hyde, both light duty men, left for the front after getting well, the nurses and Captain Nadler had a feed for them as they were such fine fellows and good helpers. The Scots sure live up to their name as fighters.

Canada has a lot of Blackfeet Indians over here with them and they are some fighters. We hear that the mixed French and Indian bloods are the fiercest fighters we have over here. It is said they don't take prisoners alive.

Klein, as noted earlier, was one of BH 12's enlisted personnel, but nevertheless reported censored feature stories home to the Chicago Daily News on a regular basis during his Army service.

Chicago Daily News, early July 1917

CHICAGO IN FRANCE HAS GLORIOUS 4TH

Base Hospital Boys Clean Up on Food and Festivities with "Tommies"

By Norman Klein

Special Correspondence of The Daily News

Chicago Base Hospital, Somewhere in France, July 4—For the first time in history American and British soldiers, comrades in war, side by side on foreign soil, celebrated together the declaration of independence to-day in a gala holiday, in the camp of the Chicago base hospital. And the event of this Fourth, which made up for the absence of fireworks and flamboyant political speeches, was the big feed in the evening, when 600 British Tommies and Chicago "sojer lads" sat down, elbow to elbow, and sang and frolicked and ate. 'Twas a grand and glorious day...

Letter #16

Laura to Erle:
France
July 5, 1917

Dearest Buddy:

. . . I hardly know what to write about anymore. Today one of the men in my ward had such a terrific hemorrhage and I'm so worried about him. It was the second he has had but this is by far worse than the other and tonight he looks so badly. He had just a slight wound but it happened to be near a large artery and so when the tissues sloughed a little (practically everything is infected) the vessel wall gave way. He has been such a nice patient—but I'm sure I've told you before how well I like the Tommies. We've also had a lot of Canadians here and I like them also. They seem about like Americans. And I like the Australians and New Zealanders, too, so after all I'm not so partial.

The Australians have the best looking uniforms and the Tommies the worst. The British officers have very good looking outfits. In fact we never see anything but uniforms. The nurses are not allowed to wear a thing but uniforms any place.

What will my one blue serge look like when I reach Chicago again!! And my hat!! Ye gods!!!

Yes, I think you & Bertha are getting up quite a flirtation. I suppose I'd better write my mother to see if she can't stop it. I'm all jealous. She wrote me also about your exchange of notes. I'm sure she and father and mother also did appreciate your sending the clippings.

She also wrote about the cyclone. It simply ruined the greenhouses and uprooted all our nice shade trees and almost all the fruit trees, too. [***Laura's oldest brother Warren established the first greenhouse, or nursery, in Jennings County, which stood for more than a century.***] But no one was hurt so after all they are lucky that it was no worse. They could so easily have been injured. . . .

Did I write you about our visit to the Red Cross train? I probably did but I can't expect all my correspondence to get through.

I think I've gotten most of mine now. And I'll have to give you credit for being the most faithful writer. I've had but two from home, two from Helen, one each from a variety of people and four from you. Your latest one was dated June 4th and Bertha's latest June 11th so I'll be expecting one soon again from you.

And honey, if you could see the way I live on letters here! I've worn them all dog-eared and crumply. I'm afraid I weary my roommate by my frequent rereadings. She probably thinks I'm a nut. And I'm not, am I?

We have heard it rumored that there are four more American Hospital Units in London but we hear so many things that I don't believe anything unless I know it's true.

There's one thing I firmly believe. And that is that the war will not last over a year. I don't know whether it's just a kind of faith in Miss Hazard's prophecies or what but I just can't see it any other way. And I do wish that it would be over much sooner.

Later.

The wind became so cool that we came down the hill and settled down for the night. It's time for lights out but I must finish while I can.

Mac just came in and said she had been over to my ward and Rankin is hemorrhaging again. Poor man. And he is so brave. If I had a hundredth of what most of these boys have I'd feel it my special privilege to howl whenever I could find an opportunity. But they shut their teeth and bear it.

The Chaplain is our official censor but the green envelope which we are allowed each week lets it out of camp without being read [***a misbelief: the envelope had a censor's seal and signature***]. Someway I prefer an absolute stranger that I've never seen and never will see to read my more intimate messages. . . .

Pvt. Baker's memoir:

July 6

I got my first half day off so Gil Read and I went to Paris-Plage on a pass. To get there we had to walk 4-1/2 miles to Etaples, a big military training center bordering an old French fishing town. Between Camiers and Etaples there is a large recruiting training grounds. They have wonderful trenches dug all over the place and go through real warfare in daily drill. They have sand to march on and it is a terror to all recruits. There are air bombardment, bayonet and wire training areas. Discipline is the word here and English discipline is wonderful, but hard. The boys are mostly from eighteen to twenty years old. They are all glad to see Yanks over here, I can assure you.

From Etaples we took a street car, very quaint old affairs, some are closed and some open, about half as wide and half as long as ours. It is about six miles to Paris-Plage from Etaples. The ride is through beautiful young forests or woods.

Diary of Dr. Neal:

July 6—Queen Mary, with very little pomp displayed, traveled through the hospital area at 3 p.m.

July 8—"Old Glory" was put up for the first time at our post, and it is to ever fly on the flag pole.

Letter #17

Laura to Erle:
France
July 10, 1917

Dear Buddy:

. . . Miss Cleveland and I have just been for a walk and it was a delight! The only drawbacks are that I can't explain just where we are and that you aren't here to enjoy it also.

I can imagine that the winters are cruel here especially with the present housing conditions but the summers are beautiful. The houses are very picturesque but very dirty. They are so different from the neat little English ones. And the farms are sloppy and smelly around the buildings but as I said before very picturesque. I've taken no long walks so far but I'm going to start in and see what I can do.

There is no lack of males here. In fact the lack is on the other side. The men are so eager to talk that the only difficulty is to keep them at a distance. They've been up the line [***at the front, or battle line***], you know, and perhaps haven't seen an English speaking woman for months. But you see it isn't good form to talk to and walk out with everyone you meet on the way.

Wednesday

Today I walked down to the beach. It is rather a nice walk but not nearly so inspiring as the hills. The sand is so fine and smooth and clean—very different from Lake

Michigan sand. We came back past a restaurant and asked for strawberries. The waitress, "No strawberries, Madamoiselle, but prunes." So we ordered coffee and bread and butter and were ages getting it. Then we found we were too late for a lecture posted at the Y.M.C.A. and learned when we finally reached home that we had missed an extra good supper here with red raspberries for dessert!!!! Such is life in a large city!!

We are trying to get our rooms fixed up with a little drapery so they don't look so bare. But sewing by hand is tedius and I'd so much rather walk than work when I have spare time. . . .

July 13

After supper Daisy and I climbed the hill again and sat and looked out over the camp and the green fields and the sand dunes.

There is one small church in the village and the bell is quite musical in the distance. So many things here are just like old paintings—I'm not referring now to the tone of the bell—and I do so wish you could see them all. . . .

Diary of Dr. Neal:

July 15—Four Belgian princesses appeared on scene.

The envelope and letter were on Oxford heavy stock stationery with a British one-penny stamp attached, sealed with OPENED BY CENSOR tape and signed in pencil on front, J.S. Dancy—the BH 12 chaplain. Letters from Laura's hometown, North Vernon, Indiana, seemed to come quicker to France than those from Chicago, perhaps because North Vernon was a small town rail center with quicker local handling of mail than the big city.

Letter #18

Laura to Erle:
France
July 18, 1917

Dearest Buddy:

. . . [Regarding press reports of the June 1, 1917, duel of the Mongolia with a German submarine] **The** write up in the Tribune of June 6th by Floyd Gibbon concerning the interview with Major Besley was correct, I think. "Hundreds of shots" were not exchanged but a few were and it was exciting enough for a while. But I didn't see the submarine and that is the disappointment of my life up-to-date.

. . . I was quite discouraged when I got your letter—probably because of the rain and because it is inspection day and the Major scolded about the men's boots not being clean. . . .

. . . You'll be terribly disappointed if your unit isn't called but really I don't think most of our boys are enjoying themselves as thoroly as they had expected. And after the voyage and the settling is over there isn't much excitement. The work becomes more or less monotonous, especially for them. Most of them have been used to very little manual labor but they've taken hold of the scrubbing, etc. manfully. I've two such nice lads in my ward. They work so hard and are so agreeable thru it all.

My work is very pleasant and not hard so far. I worked fully twice or three times as hard down on California Ave. and had much longer hours. We work from 7:30 AM to

This was one of our greatest thrills in France—10 Americans among 50,000 Allied soldiers. The whole ten of us marched off the field with our flag, passing the thousands of troops who were glad to see us, the first U. S. troops to take part in a major athletic event since America had entered the war.

8 PM with three hours off in the P.M. That makes a very nice day, I think. The privates work from 7 AM to 9 PM with three hours off when they can get it—usually they get it.

We walk a great deal here—usually down toward the beach or up in the hills. The view from the latter is wonderful. I wish you could see it. You just couldn't stay discouraged if you'd sit and look out over the camp to the sand dunes and water and wonderful clouds. . . .

I would so like to know if Units 11, 13 and 14 [***referring to other Chicago-area Base Hospital units***] have been ordered out and where they'll come to. I wish & hope they'll be some place close to us but it's hardly probable. France is a fairly large place, you know. . . .

Emma Matzen [***wounded in shipboard gun accident May 20***] didn't get to go home. They made her stay in a government hospital in Washington, D.C. till she can come on to us. She was quite disappointed at not getting to go on to Nebraska. I hope she comes soon to us. She is a dear and we all miss her.

I'm sure I've told you before of the many men here. Almost every time we go out we make fresh acquaintances who insist upon talking and walking and making "dates." It's really funny sometimes. It sounds terribly irregular but situated as we are we talk to most any one without introductions, or at best, flimsy excuses of introductions. Please don't be shocked. . . .

Pvt. Baker's memoir:

July 22

At the field day affair today, there were about 50,000 troops present, with about 400 athletes from many nations. Murphy, Strickler, Hicks, Creedon, Sweet, Kidd, Greeley of Harvard, two Boston men and myself composed the American team and we took two first's, five second's and one third. Our nearest competitor got two first's. We were given a great welcome by all the Allied troops who were here from all over the world, the U.S.A. being very prominent. Cups and money (135 francs, I believe) were given out to us by a British General.

This was one of our greatest thrills in France—10 Americans among 50,000 Allied soldiers. The whole ten of us marched off the field with our flag, passing the thousands of troops who were glad to see us, the first U. S. troops to take part in a major athletic event since America had entered the war. Individual congratulations such as "That's the stuff to give him", etc., were commonly heard. They had to sit up and take notice and they believe our enthusiasm will help win the war, which is certainly true. They were glad we won the Meet.

We went the four miles to a hotel at Etaples by "Shank's mare", as it was the only method available for traveling. We had a fine feed. The menu consisted of soup, bread and butter, potatoes, beefsteak, gravy, creamed lettuce, chicken and plums, all for 5 fr. or $1.00, and this was refunded.

We returned about 11:00 p.m. on the train to camp. There was a big convoy of wounded on hand and another came at 12:15 a.m. until 2:00 a.m. We were excused on account of being in the track and field meet. The nurse was peeved the next day because I was off and she razzed me all day.

Letter #19

Laura to Erle
France
July 23, 1917

Dear Buddy:

Do you see by the tone of my pen that I'm simply boiling? Well, I am. I went out by myself to read a new magazine (Scribner's July). An English officer came up. "I hope I'm not intruding. My pal's gone up the line again and I'm lost without him. May I talk to you a while?" And I talked and he talked.

But he doesn't like the Americans and gave us several jibes that I didn't approve but didn't know just what to say. I hate to talk to them. They're so conceited —not all but the majority are so. And then when I started home to supper he insisted upon me staying out to dine with him. And all I could say was "No thanks!" when I really wanted to biff him with my magazine. At times it's hard to pretend to be a lady. I wish every day that I were a man so I could do just some of the things I'd like to. . . .

We are not allowed to talk to the privates in our unit—nor in any in fact. We have the standing of officers (honorary of course) and army discipline is very strict and strange to us. But we're gradually learning.

The surroundings are very good to look at here. There are very few trees but I don't miss them terribly as the landscape is excellent without them. . . .

July 25th

Emma Matzen arrived this week and we are all so glad to see her. She had a shorter and less eventful voyage than we had.

The days roll by so pleasantly. We walk out almost every day and occasionally go to a concert or to a tea. But I do miss you to walk with me. You'd so enjoy the view from the hills. Last night we were on the very top. The air was just slightly misty and everything was so still. The fishing vessels were scattered along the water as far as we could see, the clouds were gray and white and the green fields near the foot are so very green. I simply can't go to the top of the hill without getting two or three big thrills. I like it very much here and I'm almost happy. Of course there are several very large "ifs" in the way.

Helen Kelley will know where we are located here. If you ever see her ask her. Miss Weil, down at 31st & California, was near here a year ago so she told Helen and Helen told Miss Cleveland's sister. . . .

Diary of Dr. Neal:

July 28—Orders to clear hospital preparatory to "big push."

There are so many horrible parts to a war. I simply couldn't realize before I came what it would mean to a nation and I have a fair sized imagination, too.

<u>*Letter #20*</u>

Laura to Erle
Appended upside down at top of first page:
We all enjoy the clippings you send. Many thanks
July 28, 1917

My dear, dear Buddy:

. . . Yesterday I had an afternoon and Miss Cleveland and I walked to the third village from here and back by the beach. I'm sure we walked twelve or fifteen miles and arrived home at 10 P.M. each with a blistered heel and a weary body. . . .

Later

I had to stop and go to supper, then Bess wanted me to go to a concert at the Y.M.C.A. And I went and was more or less disgusted at some of the raw stunts but I did meet a nice little Scotch officer who walked home with me and didn't try to make an engagement for tomorrow but was just nice and friendly. He didn't seem conceited either. Now don't imagine that I'm being a belle. I'm not at all. It's just the fact that there are so few women here and the men are so eager to talk to any one that isn't a man. We just expect now when we go for a walk to have some officer come up and talk to us. They mean nothing by it and neither do we.

We are having a convoy tonight and we'll probably evacuate all that are signed up to go. We keep shifting patients all the time. Tonight we'll send on to Blighty a young New Zealander who has been in six weeks. That is a terribly long time to keep a patient here but he has been too sick to travel before. He has lost a leg and probably the use of one hand and he's just twenty one.

There are so many horrible parts to a war. I simply couldn't realize before I came what it would mean to a nation and I have a fair sized imagination, too.

I heard today that the wards will probably be in huts before winter. I hope so for I can imagine giving baths and doing dressings in a tent in the dead of winter.

I have heard no favorable comment about our chief's letter in the Tribune asking for sympathy for the doctors & officers who must live in tents. Everyone else seems to think that it wont hurt them if it doesn't the privates nor the English officers. I have a nice M. O. (Medical Officer) Capt. Krost in my ward but the really nice ones or intelligent ones or interesting ones are far between here. Dr. B. has been quite a disappointment to me. I think I'll have to go back to my first impression of him after all.

But I've found some that I didn't like at first to be very pleasant and efficient.

Did I tell you that our chaplain censors our letters? The green envelopes go to the base to be looked over but he does the rest. Miss Urch refused to have anything to do with them.

I think the chaplain is unhappy here. He is not busy as most of the men are Church of England and he is a perfectly good Methodist. He is a pleasant chap (note my English please) and I like him very much. It is some consolation when I can't get a green envelope to know that he has a very poor memory for names. . . .

Diary of Dr. Neal:

July 29—Rain. "Big Push" is on, heavy firing in Ypres sector.

July 30—Rain pours.

July 31—Rain pours.

Aug 1—Rain pours—convoy gave us 1,000 new patients from Ypres sector.

Letter #21

Laura to Erle
France
Aug. 1, 1917

Dear Buddy:

Rain and rain and rain but I don't mind at all personally. We walked over to the village yesterday and I think I'll go out again after supper this evening. The mist and clouds are around the top of the hills but they aren't at all inviting. I'll enjoy that from a distance.

But the poor chaps in the trenches are not enjoying the drizzle, I'm sure. It must be wretched for them. But they are the most uncomplaining lot I've seen. I'm sure I'm happier in this work than I've been in any kind for ages and ages. I loved contagious work but we had so much trouble getting supplies and nurses there that much of the joy was taken out of it for me at the City Hospital.

Here, we have a weekly inspection that discourages me at times but on the whole we get thru it pretty well.

Don't you love to hear the rain on the roof? I do and ours is waterproof—they all aren't free from leaks. . . .

We've been so busy all day but I like that. I do like to feel that I'm earning my salary. But I wasn't quite thru when the night nurse came on duty and I don't like to have to leave anything over for the next fellow to finish.

Tonight the rain is still falling but not quite so hard. Imagine going for a bath and having to wear rubber boots, raincoat and rain hat over one's bath robe!! But we are delighted to have a bath house and occasionally hot water.

Did I ever draw you a plan of our huts? Here it is:

[Drawing: hut floor plan shows corridor on one side, with cubicles along opposite side of long narrow building; "Frances & Polly" in first room, "Miss C & I" in second, "Bess & Emma" in third, "Mac & Miss Stahl" in fourth, last two cubicles not labeled]

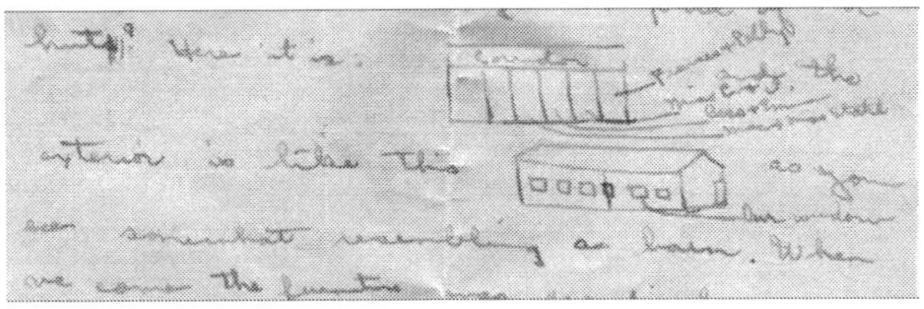
huts? Here it is:
exterior is like this as you
see somewhat resembling a barn. When
we came the furniture

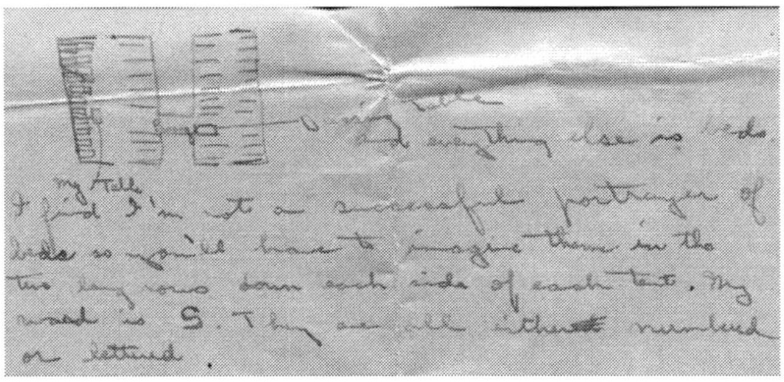
My table
And everything else is beds.
I find I'm not a successful portrayer of
beds so you'll have to imagine them in the
two long rows down each side of each tent. My
ward is S. They are all either numbered
or lettered.

Exterior is like this

[Drawing: angular view of rectangular building with gable ends, door on far side of gable end, second window labeled "Our window"]

as you see somewhat resembling a barn. When we came the furniture was exceedingly meagre consisting of two "trestles" which are two boards about six inches from the floor with a small thin bumpy tick.

Since then we've been given regular Army cots (canvas) and we've accumulated [***unreadable:**____ **ea***], borrowed, bought and stolen a chair, a library table (made out of a box), a wash stand (also a box) and two dressing tables (boxes) and since they are draped with creton and a curtain over the clothes hooks and at the window we look quite civilized.

As the walls of the huts are but one board thick I'm not anticipating a very comfortable winter but if other people stand it and if the men survive in the trenches I see no reason why we should complain.

[Drawing: ward floor plan—H-shaped with passageway connecting toward the bottom; "My table" labeled in middle of left hand wing opposite passageway and "dressing table" at bottom of passageway in right hand wing. The ward consists of two tents joined together.]

And everything else is beds.

I find I'm not a successful portrayer of beds so you'll have to imagine them in the two long rows down each side of each tent. My ward is S. They are all either numbered or lettered. . . .

Diary of Dr. Neal:

Aug 5—First clear rainless day out of past 8. Typing hundreds of men for potential blood donors during the week.

Letter #22

Laura to Erle
Aug. 6

Dear Buddy:

. . . I'm not surprised that nothing has been cut out by the censors because I've tried to be very conservative and not tell anything I shouldn't. At first I was afraid to mention anything but the weather. Now I'm a little bolder and even speak of crops and the fashions without telling what I shouldn't.

I can't help but be so glad those movies were taken at Sir Thomas'. Everyone seems so delighted with them. Silas was so pleased when he saw me in Cincinnatti. The movie man at N.V. [***North Vernon, Indiana, Laura's hometown***] was going to try to get them there so the folks could see them. . . .

There is a concert tonight at the Y.M.C.A. and every one in the hut has gone to it except this child. This child could not go because she had letters to write and didn't want to be bothered by people about while she wrote. It's bad enough to be bothered by the censor.

I've been flattering myself that the chaplain, who censors our mail except the green envelopes, didn't know me. He seems not able to connect names and faces. But he surprised me last week by volunteering the information to a patient that I had been doing Infant Welfare Work. Puzzle—Does he know me? Or was he guessing? I hope it was the latter. Anyway I have one green envelope per week.

Yesterday I had an afternoon [off]. Instead of staying home and using my pen I went to town with Frances, Pawly & Miss C. We just wandered around, had tea at a funny little nice tea room, sat on the beach, took the late tram and missed the lorrie home. So we went into a restaurant for supper and walked the four or five miles home. We had a delightful walk even tho it was a bit dusty and we hurried to get in before too late. It was quite dark at nine thirty when we reached the quarters. Two months ago it was light till half past ten. . . .

I'm sure I've told you before that I like the work here very, very much. I'm enjoying the work and the patients, too. They are most of them such nice lads, it's a pleasure to care for them. Miss U. thinks if we stay till spring we'll build our own hospital. I don't know, I'm sure, whether I'd like it much better or not. I'd like to be in an all-American hospital, of course, but I couldn't like the patients any better.

Now, please write me as soon as you know about the publicity work. I know you will. Would that I were a Mystic and could know such things without being told! I fear I'm very, very common clay. . . .

There are such oodles of places to see every place here and most of them of historic interest but they aren't comparable to our own country for living. We are so much more comfortable, I think. We don't seem to stand on our dignity so much.

Letter #23

Laura to Erle
Reverse of envelope, marked in pencil in Erle's hand:
"23 days coming"
France
Aug. 11, 1917

Dearest Buddy:

When you were a kid did you ever read stories of people sitting high on the sands, reading, writing and watching the tide come in? And didn't it sound like a fairy tale? But here I am doing that very thing. At the foot of the dune are Miss Urch, Miss Powers and Mrs. Krost. You see I'm well chaperoned and will try not to be indiscreet.

I started out alone to write letters but Daisy captured me and told me she didn't consider it advisable to go alone and she dragged me, protesting, with them. Now I don't mind because the water is beautiful and the clouds wonderful. They are piling up in great fluffy white and gray mounds all around the horizon. . . .

Interruption: A little French girl came over to beg, "Penny" and I had to talk to her. The conversation was decidedly one-sided but she grins broadly to make up for lack of English. She is now planted beside me to watch each move I make. . . .

I've told you a dozen or so times how well I like the work here, haven't I? But really it has been so long since I've really enjoyed my work that I can't help telling folks about it. Miss Urch asked me this afternoon what I wanted to do next and I didn't know what to say. I'd like some operating room work but I've had very little surgical experience as far as scrubbing up is concerned. And then I'd like a chance to go to the Casualty Clearing Station right back of the lines. But I'm just awaiting the course of events and am willing to take whatever is handed out. Miss U. seems to think I manage my ward all O.K. but I'd really like to not have the responsibility. I hate responsibility.

I've become quite addicted to the tea habit. When I miss afternoon tea I'm so hungry before supper time. I'll now be able to join you at tea drinking with a much better grace. The coffee we get is miserable but I still persist in buying it when we go to town. I keep hoping that I'll get some good if I keep on trying. And need I say that I have not tasted the French wines. But you see this is practically an English place. . . .

Later.

After supper we turned on the victrola and danced in the sitting room of the mess. One was a Katrinka record but it didn't sound much like the Jass band at Rienzi's. . . .

It's two months tonight since we landed here. If the rest of the time goes as rapidly and as pleasantly I'll be glad. But I do not anticipate the winter time with much pleasure.

We've never seen the Paris "Tribune". Perhaps it just doesn't penetrate so far as we are. But if we do get one I'll be sure to send a copy. You may be able to purchase one on the streets of Paris before a great while. I hope so at any rate.

Miss Powell, the artist at 453 Belmont Ave., writes us that a quarrelsome couple are occupying our old flat. We certainly lead a free and easy existence there. Conny and Sunny have taken a room with Miss Powell but at latest report Conny was still visiting home folks and hadn't settled down to the sordid work of amassing a fortune. . . .

I do enjoy the clippings you send. It keeps me in touch with things. We get a good many Tribunes and a few magazines thru a London agency. The Sat. Eve. Post would be very welcome if you sent it. There has been a lot of trouble about magazines coming thru. Miss U. subscribed for Atlantic Monthly and another I forget which and neither has showed up. . . .

Letter #24

Laura to Erle
France
Aug 12 1917

. . . The evenings are getting so much shorter. It is dark at nine o'clock now. We have had no hot weather and not many days when we ventured out without capes or sweaters. Of course our capes are very striking and that may account for us wearing them so much. . . .

Friday, the 14th

We are looking for our reinforcements and we may have to be three in a room instead of two. Then I'll write more disjointed letters than ever. . . .

A dark cloud is coming up so I'm afraid I shall get no farther than the post office altho I love to walk in the rain. Did I tell you that we now have rubber boots and rain hats and we certainly work our rain coats hard! I'm glad I didn't leave mine at home as I planned to at one time.

I received a large box of candy from Helen K. yesterday. It came thru very nicely. We get plenty of candy here and that is contrary to our expectations. There are certain days in the towns when we cannot buy meat, cake nor sweets as they call candy. But we can always buy candy at the canteen.

We all had loaded up on sweet chocolate and toilet soap before leaving Chicago. Now I wish I hadn't and had put in more woolen underwear and stockings. In fact I'd have packed quite differently had I known how things would be. . . .

Miss Urch is such a dear. I'm surely glad she is in charge of our unit instead of some of the old birds that might have been sent. She has her troubles.

I've told you in at least two former letters that we have not yet seen the Tribune published in Paris but we get the Chicago Tribune fairly regularly—a copy or so week—thru a London agency. We get a little N.Y. Herald published in France but it doesn't seem to amount to much.

Day after tomorrow will be the third month since we left Chicago and that's one-fourth of a year. And I've known some three weeks that seemed so terribly long. I wonder will it be three more months or a year. I absolutely refuse to consider the possibility of this beastly war lasting longer than another year. . . .

I've become quite addicted to the tea habit. When I miss afternoon tea I'm so hungry before supper time. I'll now be able to join you at tea drinking with a much better grace. The coffee we get is miserable but I still persist in buying it when we go to town. I keep hoping that I'll get some good if I keep on trying. And need I say that I have not tasted the French wines. But you see this is practically an English place. . . .

Close up of Laura and others with Sir Thomas Lipton

Letter received by Laura from Sir Thomas Lipton of Lipton Tea,
Apparently in response to a "thank you" note she had sent him

Osidge,
Southgate,
Middlesex
Telephone No. 3 Finchley
Nearest Railway Station,
Palmers Green, G.N.R.

14th August 1917.

Dear Sister Huckleberry,

I was much pleased to receive your kind letter and only wish I could see the moving picture you mention. Perhaps when I get over to America it will still be running.

I thought I would have been able to visit you all in France before now but trust this is only a pleasure deferred. I am sorry the photographs taken at "Osidge" have been so long being completed but I am pleased to say that they are now in the hands of the Censor being stamped for transmission abroad, and I hope to be able to send them on to my good friend Miss Urch in a few days. I will put the name on each and sign all the photographs.

Kindly remember me to all the sisters of the Chicago unit. I can only say that the patients under their care are very fortunate and are to be congratulated.

With my best Wishes

Yours Sincerely

Thomas J. Lipton

Miss Laura Huckleberry

U.S. Base Hospital No. 12,

General Hospital 18,

B. E. F.

France

Pvt. Baker's memoir:

August 20

We took Major McGee to the depot and gave him a rousing farewell as he was ordered to some other post. We sang Chicago, Northwestern and Illinois college songs. He was a good soldier and well liked by everyone. He was a regular Army man.

Letter from Erle to Laura's sister:

Editorial Rooms

Western Newspaper Union

Newspaper Service

210 So. Desplaines St.

Chicago, August 20, 1917.

Dear Miss Bertha,

Here is a clipping* that may interest you. Its contents perhaps will seem rather terrifying at the first reading—the statement, in particular, that the doctors and nurses are working on the battle line—but actually there is little cause for alarm over the safety of the Red Cross girl in your family.

One of our medical officers who has been in Europe within the last six months told me that during a great advance "surgical teams" (each composed of surgeon, nurse, anesthetist and private) go forward from the base hospitals to the field hospitals to help give secondary aid (as distinguished from the meager first aid given in trench dressing stations) to the flood of wounded coming back from the front. Field hospitals are several miles behind the firing line—fifteen to twenty-five miles. Hospital Unit No. 12, I was further informed, is 70 miles behind the British lines and three miles from the sea, near the town of Etaples. If you have a map of France, maybe you can locate Laura approximately. . . .

Sincerely,

J Erle Davis

P.S.—I saw Laura in the Hearst Pathe weekly news pictures at a motion picture theater recently—right up close, face to face. The reel was made in London.

* *missing*

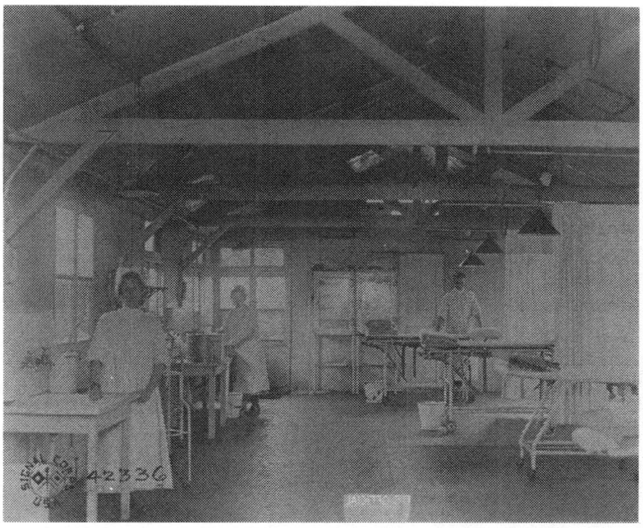

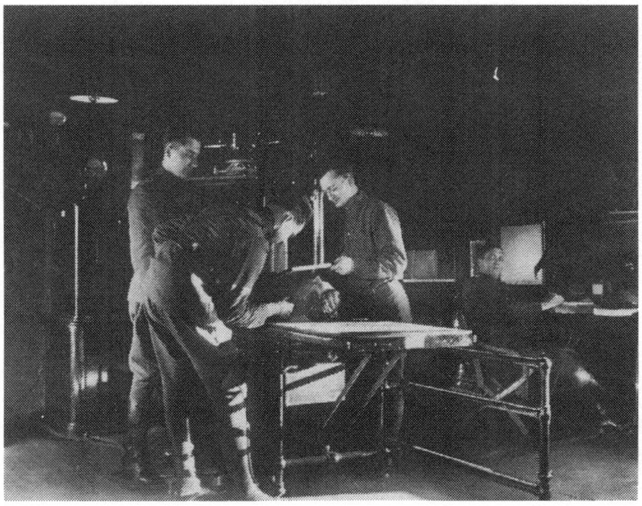

BH 12's Operating Room, paid for by Chicago Red Cross.

Men's Ward: steel Quonset-type huts erected for wards after the Sept 1917 bomb damage.

X-ray procedure.

Letter #25

Laura to Erle
France
Aug. 22, 1917

Dear Buddy:

. . . This two weeks has been unusually long because I was so eager to hear of your appointment to the publicity work and I was so sure it would come along in this letter. But it didn't. Again I'll have to wait on the mail.

Neither have I heard from home for twelve days. But yesterday I had a letter from Miss Stodola which I wasn't very pleased to get. Why, oh, why does that lady cling so closely? . . .

Tonight we have a lecture at the Y.M.C.A. by Sherwood Eddy. I heard him last Wed. night also and enjoyed him thoroly. He certainly lectures very differently from the Englishmen we've been hearing.

And Sat. night we had such a funny concert in one of the wards. It was perfectly informal and everybody who went had a good time. An old Scotchman brot his bagpipe along when he was wounded and he gave the concert. Naturally the music was barbarous—worse than a Jass band—but it wasn't music I want to hear. Then, four of the American privates, who sing together frequently, came in, sang all they knew and then a bunch of English and Scotch orderlies and light duty men, who had been hanging around just inside the door, broke in with some of their songs. An Englishman (a patient) sang a solo and all the rest of us joined in all the choruses. Between times the little Scotchman or "Jock" as we call them here, sat up on the edge of his bed and piped for us.

Now all this occurred in a ward and all of us, entertainers and entertained, sat about on the edges and foot of the beds. . . .

Yesterday I had a P.M. [off] but I didn't take a walk. Is not that marvelous? Instead I went to town with firm intention of getting something good to eat. Now, I'm not complaining of the food for it's much better than I had expected. But at times I must have something different. So Emma Matzen, Bess and I went to town.

We had coffee and toast, the toast being the treat. Then we wandered around town and looked at all kinds of fancy work and took a tram homeward, went in the corner restaurant, had steak (the first since we left Chicago), potatoe chips and pears. Then we took the lorrie homeward again. . . .

We are now to have a whole day once a month. . . .

After supper.

The lecture was very good—about Russia mostly and some about China. He'd make a very good man for the Sunday Evening Club at Orchestra Hall. He's probably been there but I've never heard him. . . .

Don't ever imagine that it would be pleasant to live in Sunny France. I have never yet heard a good word for the climate. And then it's so dirty. That little editorial from the Trib. that you sent me some time ago about the Sammies signing a promise to return home because Canadians, Englishmen, etc., were marrying and staying in France was very funny.

I don't honestly believe that any Englishmen I've met could be persuaded to stay longer than he has to and the Canadians are just as fond of their country as we are of ours.

Other places are all right to look at and to visit but the little old good old U.S.A. for mine.

Miss Cleveland is on night duty now so I'll probably go on soon and you know how I love night duty. The evenings here are so pleasant but the mornings will probably be just as nice in which to walk and dreame (I didn't mean to put an e on the dream. It was reflex action I think). . . .

Pvt. Baker's memoir:

August 23

We went to Paris-Plage with Olson, McKinsey and Dr. Dancey to secure the promise of the British Y.M.C.A. heads to erect a hut for us. We saw Mr. Scott, since Sherwood Eddy, the International Secretary, could not be there. We later heard Eddy speak three or four times. He told us about the troops in Russia and Balkans.

I saw U.S. doughboys come through here on a long train and they sure do cheer us when they see "Old Glory" flying.

We are getting many wounded troops in from Lens and Ypres. Ypres lies between the two lines and they have been fighting there for three years already—in the same spot.

Captain Dale, Major Speed and Sergeant McCash went to a Casualty Clearing Station on August 23rd.

Letter #26

Laura to Erle
Enclosed was a copy of The Chicago Tribune,
*Army Edition, Paris, Saturday, August 18, 1917.**
France
August 26, 1917

Dear Young Man:

At last the Paris edition of the Chicago Tribune has reached us. It isn't coming regularly yet but will soon and I for one am most delighted to see it. I'm enclosing one that I borrowed from Miss Urch.

Miss Cleveland's Sat. Eve. Posts are beginning to arrive now but mine has not yet come.

My last letter was a wail against registered mail. I haven't yet received the registered letter you sent on July 26th. The last unregistered letter came thru in sixteen days. I had the note telling me you had sent "Confessions of a War Correspondent" on July 28th. The book arrived in the next mail. It probably came in the same mail but wasn't delivered till later. But the registered letters haven't come.

I surely did enjoy the "Confessions of a War Correspondent." I found it so interesting I didn't let it rest till I'd finished it and Bess and Elizabeth are both reading it now and apparently enjoying it just as much as I did. Thanks awfully! . . .

It's simply pouring tonight. I do love the patter of the rain drops on the roof but it does hold back operations at the front. And the boys come in caked with mud. . . .

I've been fairly busy and happy this week. You see the patients are so cheery and pleasant that our work can't help but be pleasant. I haven't found a grouch yet and that is wonderful in a hospital.

We are thinking perhaps we'll get some of the American wounded here but of course we don't know. Will probably tell you more about that later on.

But absolutely I couldn't like the Americans better as patients. I might and of course will be partial to them as individuals.

Did I tell you that I'm taking a few French lessons? I don't expect to speak French but I want to be able to recognize it when I hear it. I have one or two lessons a week but I'm not fond of the teacher so probably I shan't continue after the twenty lessons are finished.

A couple of our doctors and one of the nurses have gone up to a Clearing Station.- you know just back of the lines. They'll probably stay six weeks or so and then someone else will go. So many are anxious to go and I doubt very much if many of us get the opportunity as we belong distinctly to this unit.

It's almost church time. I must slip on my dainty rubber boots and sally forth into the rain. I'll continue later if there is anything to say. . . .

* *The four-page tabloid-sized* Paris Tribune *for Aug. 18 presented page one stories on formation of the 42nd Rainbow Division from crack National Guard units, an unruly "suffragette" rally in Dupont Circle in Washington, D.C., and indictments for a race riot in East St. Louis, Illinois, along with editorial-page type cartoons. Inside and on the back page were other war stories from home and abroad, plus additional cartoons.*

Diary of Dr. Neal:

Aug 26—Rain and winds destroyed about ten of our tent wards, putting 500 patient beds out of commission.

Letter #27

Laura to Erle
Aug. 27, 1917 [Monday]

Dearest Buddy:

. . . The rain has been simply pouring down all day and the wind whistling around the tents. We were almost deluged in one part of the ward today. The roof ripped open last night and as we have no canvas to repair with we'll have to pray for sunshine.

The rain beats on the roof and the wall and the windows tonight and the sound of the wind makes me lonely. My room mate is on night duty and the girls in the next rooms are asleep. I've put buttons on a clean uniform, polished my shoes, washed my hands and face and am all ready for bed but haven't the inclination to retire.

After supper tonight Emma & I went for a walk down the road. We wore our rain coats, hats and rubber boots but were quite wet anyway when we reached home. A convoy was coming in. I know those lads are tickled to death to get into a nice dry bed. We're expecting another convoy before morning but it may not get it.

I have but seventeen patients in my ward tonight so have lots of room. You see the patients are continually shifting. We don't usually keep them more than a week and often not more than three days. One New Zealander was in for six weeks but that is extraordinary. . . .

Tuesday P.M.

The wind of which I wrote last night has kept right on but growing worse all the time. At breakfast I heard that the roof of S Ward [Laura's responsibility] was gone. I hurried over and found the roof off one marque (markee) I don't know how that should be spelled. The other three were intact but blowing about considerably. We moved everything down to the good end of the ward and awaited developments. But so far no further damage is done in my ward. The other wards have been going down one at a time all morning and mine is in as good or better condition than any of the others. Some are completely demolished and others only half.

Now I'm beginning to wonder what will happen. Will they rebuild the tents, put up huts, evacuate all the patients, divide us up among the various units or move us back of the Americans and put up our own hospital? ? ? ? ? ? ? ? ? ?

Nobody knows yet so it's no use to wonder.

Fortunately the hospital was running light—only 900 or so patients and we've moved them all under shelter either in the huts—we have three, you know in the church army hut or into markees that are still in good condition.

The rain has kept up pretty steadily also and it simply drives thru every crevice.

I can't help but think of the poor chaps up in the trenches. I'm afraid they'll drown if this keeps up. . . .

I forgot to say that the other tent hospitals in the neighborhood have not been damaged as ours has been. Our position is a little higher and we catch the sweep of the wind.

I'd love to walk down to the sea. I imagine the waves are quite wild. But there is nothing to break the wind for the two miles and I'm afraid I'd be blown away.

Erle to his mother, Mrs. John McAlpin Davis, in Indianapolis

Editorial Rooms

Western Newspaper Union

210 So. Desplaines St.

Chicago, August 31, 1917

Dear Mamma,

You see, I haven't gone yet. There has been another delay somewhere and the date of our departure has been set back indefinitely.

Tomorrow the unit will go to Lake Geneva, Wis., ninety miles away, by automobile, and stay there at the barracks over Sunday and Monday. They tell us it is a practice run; as a matter of truth it will be about as much a social affair as anything else. A number of wealthy Chicago people, interested in our organization, have summer homes at Geneva. We are to be their guest in a manner, and for the treat will give an exhibition of what we can do in the way of drill movements and first aid practice. I anticipate a pleasant excursion.

Jack Rees [*Erle's cousin, a midshipman at the U.S. Naval Academy*] wired me yesterday, saying he wanted to spend this coming Sunday in Chicago. I asked him to postpone the visit a week, and had a message this morning telling me he would do so.

Enclosed is both sides of the family history set down in substantial form—a copy for each of the girls and one for you and Dad. Will you ask Dad for some facts about his mother and her family. I forgot to ask him the other day, and would like the information.

And will you tell Dad, too, that I'm sorry I forgot to wire him on his birthday. I fully intended doing it, but was rushed with work yesterday and the intention slipped my mind.

There just isn't any other news.

Much love, Sonny

Arrival of ambulance "convoy"

Letter #28

Laura to Erle
France
Sept. 2, 1917 [Sunday]

Dearest Buddy

. . . Your meeting with the Palmist and her alleged message was funny. I had a good laugh to myself. The work is not horrible altho war is and I've never been so happy in my work as I've been here. The Tommies are ideal patients even when badly wounded and I think that is why I like it here so well. All other conditions are not ideal but they are fair and much better than any one could reasonably expect.

As far as the unit being "worn out from overwork" is concerned that is pure bunk. The doctors and sisters (as we are called here) are certainly not overworked. I'm sure most of us have been working much harder at home. The privates have a worse time because the majority of them are unused to manual labor as this is and are often on guard or called out for convoys at night. And we do need more men. But no one is suffering from overwork I'm sure. Please don't believe all the newspapers say—nor even half.

Of course at times we are rushed but more times we aren't. And I suppose it will always be so. When a push is on there are naturally many more patients than at other times. The hospitals must be prepared at all times to care for the push but we seldom are filled to capacity.

The letter you enclosed from one of our nurses was fairly accurate but I'm glad glad glad that you didn't take up her suggestion about the milk chocolate. I'm completely fed up on it. And besides that it is obtainable at every shop and canteen.

Of course the prices are higher than at home but we can expect that.

Fresh fruits are at a premium here but we get some fairly often. The prices are exorbitant but neither do we mind that so much. There is but little one can spend one's salary for anyway. . . .

Probably only a few of us will have the opportunity to go to the Clearing Stations. Not many are needed there and it is considered a privilege to go even tho it is very dangerous work. One of our nurses, two doctors and an orderly are in a Clearing Station now.

That very foolish writeup in the American was a scream. Why we are in the neighborhood of 40 or 50 miles from the line. Occasionally if a push is on and the wind is just right we can hear the big guns faintly.

Afternoon.

Yes I'm writing this on duty but as before stated I'm not always busy. . . .

In my last letter I told of the wind storm that wrecked so many of our tents. We are very nearly established again. They set to work immediately and put up other tents and re-roofed where necessary. I think it was a very good thing that it happened now instead of waiting for cold weather. They say the wind is very bad here in the winters. Ooh! And I hate winds so. But one can't always be a butterfly and enjoy life.

We haven't had a convoy since the storm so that accounts for the lightness of the work now. . . .

Air Raids, Troop Mutinies and Routine

Pvt. Baker's memoir:

September 2

Sunday night at 9:00 p.m., a convoy came in so we missed church. At 10:15 p.m., when we had just retired thinking how beautiful a night it was for an air raid, we had just hit the bunk when—bang! bang! boom!, etc., etc. Things began to look exciting as there was the much expected air raid on Boulogne, fourteen miles north of us. The huge searchlights pierced the sky looking for the night air raiders. We could see the rockets sent up by the gunners and we could see the star shells and tracer bullets gliding all over the sky trying to locate the Boche's machines. The antiaircraft guns boomed for twenty minutes or so but the objects—Hun planes—were not even seen. They were too high. The Huns dropped a half dozen bombs but none did much damage, although there were six killed, it is said.

September 4

At 10:00 a.m., a Fritz flew directly over our camp taking photos. He was so high he could not be seen with the naked eye. All the antiaircraft guns opened up on him but all we could see were the steaks of white puffs from the bursting shells all the way to Boulogne (north). The objective of taking photos was well carried out in spite of all the guns and we all expected a raid that night because the moon was full and everything was set for a raid—perfect weather, too.

We retired about 10:00 p.m. The moon was coming up with a great ray and the sky was clear. I did not go to sleep because there were some of our boys out talking about an expected raid. About 10:45 I got up to see the rockets and hear the shells burst up north over Boulogne. While we were all straining our eyes looking the sky over, we heard some bombs drop about four miles away over Etaples, the great hospital and training camp. Then all was quiet and we could hear a motor humming. It sounded nearer and nearer and seemed directly overhead as were straining our eyes to locate it. All of a sudden I heard a zi—zzzzzz—rrr, then saw a streak of fire, evidently from the time fuse, come tearing down about 175 yards away, when BOOM! she went and there was an awful explosion and concussion, affecting one's ears. I sure thought my seconds were counted. Dirt and pieces of shell flew in all directions.

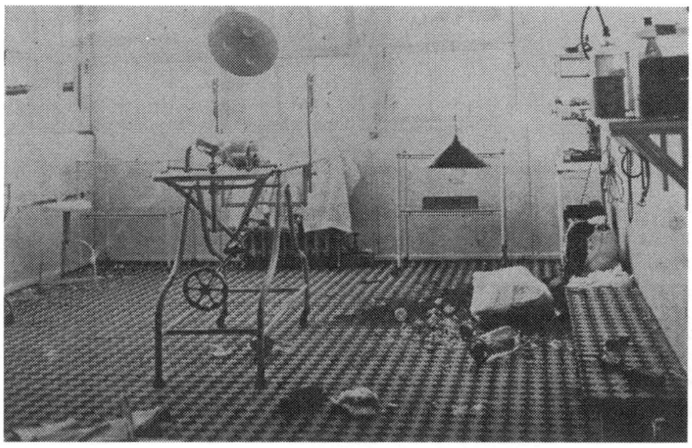

Bombing aftermath.

Opertaing room, after the bombing.

Exterior of bombed Base Hospital No.5.

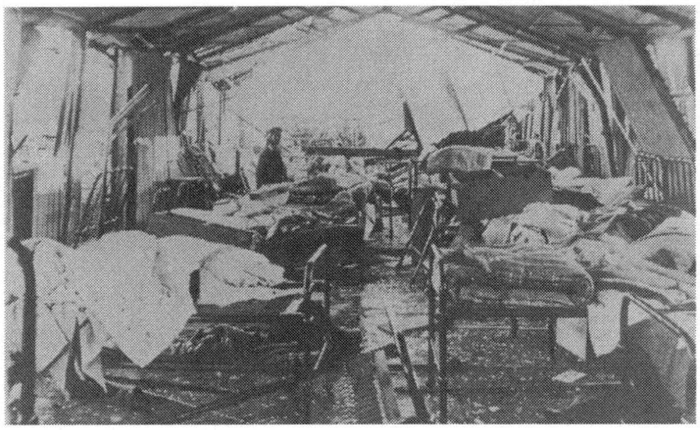

Etaples after the bombing.

Interior of Base Hospital No.5.

We were all speechless. I was standing with some others about twenty feet from our tent. We did not know what to do so we all started for the big dugout way out on the hills, but it was too far.

Immediately after the bomb hit at the end of our camp, the plane tore diagonally over toward the Boston Hospital #11. A bomb fell in Hospital #4, a dud, five fell in Boston Hospital and two of these on the outside of a U.S. lieutenant's tent, blowing him to pieces. It was Lieutenant Fitzsimmons of St. Louis who got killed. The other three bombs killed a bed patient and three U.S. soldiers, one of whom was Lester Woods of Streator, Illinois, a bugler. Only an hour before he had been talking to four of us about a "V" hut.

I thought the whole place was blown up. We went down to see the damage done in our wards and found a hole four feet deep by ten feet across, like a bowl. The head and nose cap went on down a few feet further. The English engineers dug it out. Shrapnel blew all over and two patients were wounded. There were many holes in all the tents around. There was some "shirt tail" or "pajama parade" after the first bomb exploded as every one was so scared and excited. A few minutes later we heard heavy firing and saw the shells bursting over Boulogne.

The Boston Unit officers, our officers and some other men, among whom was my tent buddy, were the Guard of Honor at the funeral of all the U.S. killed at Etaples, where a very interesting but solemn service was held. Two Scotch bag pipers played and Hanish, our bugler, sounded Taps.

These men that were killed were the first Americans killed in France since U.S. declared war in April, 1917.

Diary of Dr. Neal:

Sept 4—Without gas masks or helmets the personnel of our unit, one of the four group hospitals, received the first baptisms of fire and death. The real damage affected U. S. Base Hospital No. 5, the Boston Unit, General Hospital No. 11. Death or injury befell the first members of the A. E. F. in France, killed or injured by the enemy while in performance of duty. First Lieutenant William Fitzsimmons, officer-of-the-day, and three enlisted men were killed instantly when a bomb fell where they were grouped for the 10 p.m. guard report. My visit with the Lieut. ended less than 10 min. earlier.

Letter #29, Laura to Erle

France
Sept 8, 1917

Dear Buddy:—

I have such oodles to talk about tonight that I scarcely know where to start. I'll start on the freshest subject however.

Today I went with Miss Urch and a bunch of other people to call on an artist lady and I did have the charmingest time. You'd have loved it, too. The artist, Miss Simpson, (such a name for an artistic temperment) has a cunning little cottage on a funny winding road and Saturday is her "At home day" and there were oodles of people dropped in for tea and to look at the pictures.

She is a dainty, gray-haired person and a good hostess. She served delightful little sandwiches and cakes. But that was only the beginning. I could hardly sit still and eat I did so want to look at everything and explore the house. She has numbers of paintings some of which have been exhibited in the Salon in Paris and some beautiful etchings of both local subjects and of Venice and Algiers and a few of Ipres I simply had to have some of them and so did every body else

The house is more than two hundred years old—just a mud cottage—which she has remodeled by adding a studio and making it more livable without losing the old French lines. There is small blue tile fireplace and two other huge fireplaces with andirons. The garden is also charming but not more so than many others. It is a sort of settlement of artists or was before the war.

The girls came in just then to talk awhile and of course I had to stop and be sociable. But they completely ruined my line of thought. So I'll start another.

These lovely moon light nights have sort of lost their romance for us. I'm sure you've read about the air raids during this week and probably about our neighbor, the Boston unit. For two nights we heard the antiaircraft guns at a distance and on the third night had the real thing here. Our hospital was fortunate however. Fritz missed our farthest tent by about fifteen feet and we're very thankful he did. The Bostonians were less fortunate but I can't go into detail. They are just across from us and down the road a couple of hundred feet. And as I said before we realize that there isn't so much joy in moonlight in the war zone as in the U. S. A. . . .

These lovely moon light nights have sort of lost their romance for us. I'm sure you've read about the air raids during this week and probably about our neighbor, the Boston unit. For two nights we heard the antiaircraft guns at a distance and on the third night had the real thing here. Our hospital was fortunate however. Fritz missed our farthest tent by about fifteen feet and we're very thankful he did.

The wild tales you sent cut from the "Examiner" and "American" were pure bunk. We are a base hospital unit and do not go gallivanting up to the line at all. Who would take care of our hospital if we tagged along when there is a push on? That reporter must have been smoking more than hop.

We did send an "operating team" to clearing station but I probably will not have an opportunity to go. Not many of us will go.

I wonder if your unit has been called yet. At times I hope it isn't. I'd so much rather you got the other R.C. work. I believe it will be so much more satisfactory to you. It doesn't take brains or at least very many brains to scrub floors, carry stretchers, etc. and why waste good material on such work? I have two of the nicest boys as orderlies in my ward. One was a clerk in Carson, Pirie, Scott's [Chicago department store] and the other is an embryo lawyer. The latter is such a bright young fellow and he is a peach of an orderly. The patients are crazy about him and so are we all. But he could do so much more. He, like you, wasn't heavy enough to make the

On Sunday afternoon, September 9, there was a great uprising among the soldiers at the big Army Base at Etaples...

At present all M.P.'s are kept locked up for fear of their being killed by the troops. The Tommies demand more freedom, many want the M.P.'s abolished and they demand taking the Women's Army Auxiliary Corps girls back to England.

regular service but he is so anxious to do more. He wants to try for transfer to the officer's training camp in or near Paris but I'm afraid the commanding officer won't let him try and it needs his recommendation.

Such recommendations were refused to two others.

Did you ever think of trying for officers training camp? Or have you had enough mathematics? When I see the intelligence or rather lack of it displayed in some of the features of the commissioned officers it makes me sick to think of such perfectly good brains as yours being under such men. Of course they aren't all like that but some are. I've liked some I've met, very well indeed. I've also learned that promotion has little to do with efficiency. It usually has to do with the fraternity you and your commanding officer belonged to in college days. Of course in a medical unit only a doctor can be a commissioned officer but there are the non-commissioned officers, you know.

Probably this page will be heavily censored but I'm going to risk it. I've been wanting to say this for some time. . . .

I mustn't forget to tell you how very much I enjoy the S. E. P. I take it to my ward. The orderlies eat it raw and the patients, also, especially the young Canadians. Naturally the Englishmen prefer their dinky little English magazines. That of course is perfectly natural.

I had a young English school master (as they call them), a Cambridge student, as light duty man—that is he washed dishes, mopped floors, helped the orderlies, etc.—I took over the "Crushed Flower" and the "Confessions of a War Correspondent" to him and they were a great treat for him. He was very enthusiastic about "Judas Iscariot." We have some books in the wards but a really new book is a novelty, you know. . . .

Pvt. Baker's memoir:

September 10

On Sunday afternoon, September 9, there was a great uprising among the soldiers at the big Army Base at Etaples. Some Military Police, (Red Caps), started it by killing a Canadian "Jock" (Scot) and in a few minutes there was a wild demonstration which was resumed every evening after that after the boys got out from supper. At present all M.P.'s are kept locked up for fear of their being killed by the troops. The Tommies demand more freedom, many want the M.P.'s abolished and they demand taking the Women's Army Auxiliary Corps girls back to England. These are working girls who volunteered to serve in base areas in France to help at anything—cooking, running ambulances, etc.

I went through the grounds this morning and saw thousands and thousands of troops. Bombs, machine guns, rifle fire and anti-aircraft guns are going all the time. The troops all looked gloomy today as the trials of three years have plainly taken their toll. There are new men coming every day from Boulogne, fresh from England. Many are 18 to 20 years old.

Letter #30

Laura to Erle
On envelope reverse in Erle's hand:
"Rec'd Oct 11", indicating more than four weeks' transit time.
France
Sept. 11, 1917

Buddy, dear,

. . . Today I had morning hours. I climbed the hill to feast my eyes and to write you a letter. But I didn't succeed in doing the latter because of interruption by a little Yorkshire officer. They do make themselves a nuisance at times. . . .

Do you really think the war will be over in a year or less? I had an awful fright a few nights ago when I got to thinking that maybe it would last two years or more. That was the first real doubt I've had. War is such a hideous thing.

. . .

And then arrived by way of the nice curly haired mail carrier (one of the privates) your letter of August 23rd from Indianapolis. Tickled ?? Of course !!! And I feel sure there's one missing between this and the last and it will be coming along some day soon. I'm glad you had such a nice little visit at home. Did you just take a one week vacation this year? You should have . . .

[*Top one quarter of next sheet torn off—unclear whether from censor or a later removal; if censored, it was the only excision in the more than 150 letters. Conceivably it reported the aforementioned mutiny, which likely frightened the nurses.*]

. . . .bellum days this part of France was a sort of rendevous for artists on account of the wonderful sky effects. For once their artists and I agree.

This afternoon when I went over to the mess for tea (oh yes, my dear, I take tea every afternoon and also bread and butter and jam with it) I found the little artist there with our etchings. I'm so pleased with them. . . .

I'd like to tell you a foolish thing I did last night but it would look too idiotic on paper. I'll tell you when I see you perhaps.

[*The other half, the back side of the torn-off quarter sheet, was missing here.*]

. . .They've not molested us but. . .

Tonight

I don't know what I was going to say about "tonight" for at that moment I had to go to a patient and then I had backs to rub and dressings to change and pillows to adjust and just as I was finishing up getting all the men ready for the night a convoy arrived. And now it is quite late and I must hie me to my little bed.

Wednesday [Sept. 12]

I find that I forgot to mail this letter so I'll add a few lines.

Today I had a P.M. and I just lazed around, looked at the picture papers you sent and tried to catch up with the Sat. Eve. Post. Then after tea Emma and I walked to the sea and walked on the sand and watched the sun set behind gray purple clouds. Homeward, or possibly I'd best say campward, thru the dusk was also pleasant. . . .

We've heard today of the state of civil war in Russia and of the trouble with Sweden and consequently every one is a little blue.

Did you know that Conny is coming, in fact will be here soon? [***May Connard, former roommate from 453 Belmont who had been in Germany—see Appendices.***] They put up little bell tents today as I suppose they are in London. Several letters have come already for her. Isn't that fine to have mail waiting when one arrives? I know. . . .

Erle to his mother in Indianapolis

Chicago, Friday Night (September 1917)

Dear Mamma,

This is written in the hope that you have fully recovered from your illness of the last week or ten days and that you are all comfortably settled in the new home. I came out to the house this evening with full expectation of inditing a nice long old-fashioned gossipy letter to the family. But I've developed such a cold in the head, or have acquired hay fever or something, that my ambish has oozed away.,

It wasn't that I have any real news to impart, but I was much in the mood to visit with you by mail. We haven't any idea yet when the unit will get into action. I hope we go within the next few days (but we'll not) because Mrs. Alward is giving up her apartment and I'm about to have to hunt a new place to shelter me. Isn't that the very devil?

I want to write both the girls long letters. Betsy, I want to congratulate for her splendid work in making the change and seeing things through like a man and woman too. And Hettie, I want to tell something about newspaper work—though all she needs is just to get started and to lose self consciousness. Tell Dad I sympathize with him in that he won' t have even a furnace to spit into this coming winter.

Yesterday's papers carried a dispatch telling of Myron Bertman's death in France. Tough luck for the old scout to get over there only to have to die in bed without one single thrill—or one dead Barbarian to his credit! Tough luck indeed.

Much, much love

Sonny

P.S. Now that you have plenty of room I will send all my stuff to you, instead of sending a part to Louisville.

Letter #31

Laura to Erle
France, September 16, 1917

My dear Buddy: —

Four months ago tonight I left Chicago. Do you realize that one third of a year has passed? I only hope the other two thirds go one half as rapidly. . . .

We are quite busy at our house lately and I have [had] a beastly headache for the past day and a half and so, honey, you won't mind if I don't write anymore tonight. I'll write again later in the week, sure. . . .

[*Postscript*] You may tell Major Donne (sp?) that up to date no one's nerves have frazzled. He must have a large fund of imagination to draw from. One boy with a bad Nephritis had to go home and others have been ill but I hardly think from overwork. . . .

Letter #32

Laura to Erle
Sept. 17, 1917

Dear Buddy:

. . . When you write of your Aunt Sadie thinking I'm doing such a fine thing and Carita speaking of the "worthwhileness" of my work, I feel very, very humble.

Why, honey boy, I'm not doing any thing that any woman with the training I've had wouldn't be glad to do. It's, after all, so much a matter of chance that I happened to belong to one of the units that was called out first. And I really enjoy the work. If I hated the work and the country, etc., and yet stuck to it I might have some reason to be proud of myself for sticking. As it is it's really a pleasure to do it. I only hope that the war will soon be over. Do you think it will be finished next summer?

I've just finished reading Vernon Kellog's article that you sent. I fully agree with him that Germany must be crushed and I don't doubt that she will be eventually. The only thing that bothers me about that is the time it will take to do it. I can't bear to think of it running on indefinitely.

I hate to think of the men running the fearful risks they do physically, mentally and morally. I'm beginning to wonder what kind of citizens many of them will make when they return to civil pursuits. The large majority will be very, very glad to have the opportunity to go back to their families. I imagine the biggest problem will be with the younger men who haven't definite occupations nor strong family attachments.

Even these youngsters here with us go down hill all too rapidly if they will, and it must be infinitely worse up the line.

Yet I can't bring myself to believe in a "Peace at any price" program. . . .

I've been expecting to be put on night duty at most any time but I'm not anticipating it with much pleasure. I'm too fond of my own ward. To me it is the very pleasantest place in the hospital and I don't want to give it up. . . .

Tuesday

This afternoon I went over to the village to buy fruit and eggs. We paid about ninety cents for a dozen eggs. Do you think the French hens will starve at that rate? I don't. In fact I'm beginning to think that while France as a nation will be poorer after the war the people will not be. They are making hay while the sun shines and bleeding every one they can for all they have. But perhaps they need it. Someway it seems a rather crude thing to do but it is of course the individuals and not the nation. When I hear the men tell of either being refused drinks of water or else being charged so much a cup I can scarcely believe it. We, in America, hear such very different tales of the French and the Belgians—especially the Belgians—than we hear over here. . . .

I'm beginning to think that while France as a nation will be poorer after the war the people will not be. They are making hay while the sun shines and bleeding every one they can for all they have. But perhaps they need it. Someway it seems a rather crude thing to do but it is of course the individuals and not the nation. When I hear the men tell of either being refused drinks of water or else being charged so much a cup I can scarcely believe it.

Pvt. Baker's memoir:

September 21

Our squad, the "Famous Fighting Second", (meaning we are in the second hut of the five), was pulled out for a convoy of wounded men at 1:30 a.m. and we carried stretchers for two hours. The boys were all from Ypres, the terrible death hole for three years. Ypres is in the South West corner of Belgium. The English call it "Wipers", "Eeps", etc. It is pronounced like E'pree.

Letter #33, Laura to Erle

France
Sept. 21 1917

Dear Buddy: —

. . . Tonight we were so busy. We had quite a good sized convoy and we examined and dressed their wounds before we came off duty. I was glad to get it done because tomorrow I'll be alone with the V.A.D. in the ward and now we can go right along with the dressings without waiting for the M.D. to come to look at them. And we have thirty seven patients and may have fifty by morning.

Our additional force has not yet arrived and we'll be quite glad to see them. We've been expecting them every day but of course won't know till they appear on the scene.

The papers were certainly full of the American preparations for war. I'm getting really inspired myself and I just can hardly wait till we really get into the fighting. I'm really expecting the Americans to just naturally chase the Fritz' clear over into Russia.

I did enjoy this bunch of papers you sent. They were so full of enthusiasm and pictures and everything. The other girls seem to enjoy them also.

Yesterday Miss Cleveland and I walked down to the sea. We lay in the sand and watched the waves and the clouds and the sail boats. I carried along the "Lockerbie Book" and sort of lost myself in it. I do like Riley very very much when I'm in the mood for him. Yesterday I was in that mood.

We've been so busy today that I've come home this evening just as happy as can be—not because I'm glad to see these boys wounded but because they themselves are so tickled to get into a nice clean bed and to be all fixed up comfortably. One just can't help but enjoy them as patients.

I didn't begin this with the intention of finishing tonight. I hadn't anything to tell you. Just wanted to talk. So I'll say "Goodnight" and write some more when the spirits so direct me.

Sunday 10 P.M.

I meant this to be a nice long letter, Honey, but circumstances have interfered a bit—or a "wee bit" as our Scotch lads say. We've just been so busy the past two days that I can scarcely think when the day's work is over. We are running short of help and looking every day for our additional force —nurses, doctors and orderlies—to come. They're probably being held up some place waiting for passports, etc. I surely wish they'd come soon. . . .

I'll say goodnight now, and will write again when I can get time and am not too tired

Pvt. Baker's memoir:

September 27

I was invited to "O" ward in the evening where the boys of the jazz band and some Volunteer Aid Detachment girls and some patients gave a little musical and farcical comedy. It was very good. Miss Brown and a V.A.D. gave a recital. Miss Bloomfield, also a V.A.D., gave Miss Brown's poem on the ward. These V.A.D. are well-to-do English girls who voluntarily come to France to help the nurses. Many are Ladies and title holders. Captain Nadler, Miss Sparks and Peckham are very clever. The good time ended by old time songs, "God Save the King" and "Star Spangled Banner", after which all marched to the other marquee and enjoyed some coffee and cookies. It was 10:30 before all was over. We held another one three nights later. Captain Nadler is a regular guy. He is a Chicago doctor, full of pep and a real human being.

Letter #34

Laura to Erle
France
Sept. 29, 1917

Dear Buddy:

. . . We've been so terribly busy for ten days now that we've hardly taken a deep breath and there's no let up in sight till our reserves come and we've heard nothing from them.

You did have a nice visit or rather outing to Lake Geneva. I enjoyed your account of it and also the pictures you sent.

The practice stunt of the stretcher bearers I thot especially good. Your kodak must be an excellent one as the views are so clear.

The boys are giving a concert in one of the wards very close here and its very good. There are some very good voices and musicians among the privates. They surely find appreciative audiences.

Sunday night [***Sept. 30***]

. . . life is so strenuous hereabout that when I get into my room I haven't enough energy left to think. The only trouble with such busyness is that when I sleep I dream work and hospital all the time. But such is what we came for. It does make one feel as if we're needed. There was a while when I wondered if we hadn't better have staid at home as far as any necessity was concerned.

I have the nicest doctor [Krost] in my ward. I'm terribly fond of him. He's so good to the patients and they all like him so well and I really feel that he does all in his power for them. When you have known as many poor doctors as I have you'll appreciate a conscientious one as much as I do. His wife is along with us as a nurse. She is a graduate from Presbyterian Hospital and is a dear. And isn't it nice that they could both come?

Fritz flew over today but dropped no bombs. He rarely does in the day. He is usually just scouting. Then at night he comes with his bombs. I don't mean by that, that he comes often. So far we've had but one raid. Some of the nurses are very nervous about Fritz but they have nothing on the Medical Officers and other officers. So far I haven't been able to worry about him. It is simply a matter of whether he hits us or not and I feel that the chance of him doing so is very slight. Please don't worry. One air raid in four months (and we all escaped that one) is a fair record, don't you think so? . . .

Emma and Elizabeth have gone to church. The others, Frances, Bess and Ella are on night duty so everything is as quiet as a mouse except the little clock and a few mosquitoes that are flitting about my ears. There is a delightful autumn tang to the air today and yet is not cold.

This afternoon Miss Urch and I went back of the camp up into the hills to a quiet grassy spot, lay down in the sunshine and slept. Not very romantic but we were both so tired. I wakened hearing a man's voice across the little valley. It was a shepherd with his flock. He called to them and they crowded around him and went along just as in pictures[.] [W]e see them [***missing word***] up over the top of the hill and out of sight.]

Then away off thru the haziness we could see the sun sparkling on the sea. . . .

I've now had four S.E.P.'s, a Collier and 4 bundles of papers and I also received the three books as I'm sure you know by this time. It was very thotful of you to order coffee for me, too, but why did you do it? You know I'll begin to be afraid to talk along to you for fear I'll mention something I haven't and then you'll write and say you've started it. But honest to goodness I'll be glad to taste some honest to goodness coffee. All we get is so adulterated and mixed with chicory even tho we pay 80 cts a lb. for it that it doesn't taste like anything.

Well, my dear, the girls have come from church and camp gossip is the order of the day. I'm afraid I'll have to stop and I'm just getting a good start. . . .

Nurses quarters.

Fall 1917, October–December: War, Wards and Weather

Pvt. Baker's memoir:

October 1

We were just half through eating dinner when, Boom!, Boom!, we heard the antiaircraft guns speak and we knew instantly a Boche* was near, probably taking photos as he often does. We had to look straight up to see him. We could hear the motor and with glasses the fellows spotted him. The duds, or unexploded shells, began falling al1 around and you ought to see the fellows dig for the mess hall, or any covering. About fifty shells were fired at him in three minutes.

The next day the same thing happened except that there were two machines taking their own time, coming straight over our hospital. The cannons for five miles ahead of us had opened up on them. The sky was full of white puffs and some black puffs of smoke from bursting shells for miles. No less that forty guns were after him. Their object was to keep him high so as to spoil his photos, if any. Both sides have developed the telescopic lens which can take good photos on a clear day even though the plane is three or four miles high [*more likely two to three miles, as aircraft of the time did not have the required oxygen equipment to operate any higher*].

Well, I was helping load a big truck to go to Etaples and it was very exciting to watch the spectacle. Our truck driver got his wind up and we went on down the road, the two Germans directly over our heads. The shells and shrapnel were falling everywhere. We saw the civilians beat it to their homes and many soldiers, cooks, etc., lay in the ditch along the road so as to save themselves from flying shrapnel if any bombs were dropped. We soon fell behind and they went over the place we were headed for and then turned back. Black smoke from some shells mixed with so many white puffs, made it interesting. Duds fell all over, each weighing twenty to thirty pounds. Wow!

No country in all creation has a moon like bloody** France, it seems, and that night it came out beautifully. The night was quiet and warm. The kind of night I love to walk with a lady through the park. At nine o'clock, just after a big convoy of wounded from the Ypres front were carried in, a heavy bombardment began of Boulogne, twelve miles away. We could see the lights in the sky of the shells bursting and could hear the aerial torpedoes of the Huns roar when they hit the ground. It seemed as though a hundred guns were firing into the sky. Boulogne is the landing place for English troops. The Germans are trying to ruin it and handicap the British movement of troops. There must have been a dozen planes in the air. When the train pulled out at 9:20, the enemy planes followed it for five miles, dropping bombs all around it but made no direct hits. Some of our unit were on this train. The Hun turned back and dropped many more bombs around the railroad station and docks. We were seated on an embankment facing all the doings and it was one thrilling hour that we put in.

* *French slang for the Germans*

***Note that Baker has picked up both French and British slang.*

Letter #35

Laura to Erle
Oct 3, 1917

Dearest Buddy:—

All the afternoon I've been trying to get started on this letter to you but just haven't. You see last night we danced in our dining room till eleven. Today Miss Urch is sick so I sat with her while Miss Matzen rested up a bit. This evening we heard that our reinforcements are almost here but they haven't appeared yet and probably won't till sometime tomorrow.

Haven't we been having the most gorgeous moonlight nights? But, as I said a month ago, such nights are more fun for Fritz than for us. I must tell you a joke on myself. A couple of nights ago one of the orderlies in my ward received a bag of pop corn from home. As he had no way of popping it he asked me if I could manage it. Our little V.A.D. (a dear little half French & half English girl from Jersey Island*) had never seen pop corn so she came along with me to the kitchen. I also gathered in Miss Matzen to help. We had almost finished popping when the lights blinked the signal for an air raid and then went out. You see that's what they do over the whole district. We finished our corn by candle light, filled our little baskets and then started over towards Halls' tent with the loot.**

Every one was out looking and listening—great sport this watching for air raids—but we went on doing just as we had planned. We went past one bunch of tents past a big bunch of men, crossed the road & just as we crossed the corner something went z-z-z about 5 or 10 feet behind us. It sounded like a dud shell from the anti air craft guns. We ducked and ran. Such a loud yell and laughing!!!

One of the boys had let off a Victrola spring at our heels. And we felt like idiots of course.

The next morning I certainly hated to go to work and face those boys but I managed it all right. It seems that the one who set it off didn't see us coming but intended it for the bunch of men standing watching. We turned the corner just in time to get the full effect.

Today the young man whose genius had thot up the trick was called up before the Captain and scolded. The Captain was on the hill and saw it all and probably laughed as much as anyone else. But discipline is discipline, you know.

Of course a very few of the girls are very nervous and I suppose if they had been the unfortunate ones they'd have had a fit or something. As it happened none of the three of us are at all nervous about such things. It was simply self preservation to duck. Now listen to me explain all the fun out of it. We certainly had a good laugh at ourselves.

A couple of days ago I really saw a Fritz machine. He flew over the camp but very high up. He really looked much like a puff of smoke from the shells from the anti-air craft guns. Of course this isn't the first time he has been over but it's the first time I've been able to see him.

So much has happened this week. I had a letter from you dated Sept. 12th and also the coffee arrove. The cans were broken and crushed and everything but most of the coffee was still in the box. Whatever or whoever sat on that package was heavy and sat hard. Ooh! The delightful smell! Or rather odor!! Perhaps you'll know by the time this reaches you how chicory tastes as coffee and then you'll understand the longing I've had for real coffee.

Now, please don't send me anything else. You see I can't send you anything and I don't like too one-sided an exchange of offerings.

We didn't make coffee last night because as I said about two pages back we were entertaining friends by a dance and tonight Daisy is ill and Emma is taking care of her. Perhaps tomorrow will be our lucky day so we can have a coffee spree.

I'm surely glad that we're going to have more help. We've had so many sick patients and such heavy dressing for two weeks now that we could scarcely get thru them all running short handed.

Did I tell you that my pet orderly is an embryo lawyer and comes from 541 Barry Ave. which is now a fraternity house. He is such a fine young man and so perfectly likable. He's wild to go up the line. But he wants the chance to take examination for commission first.

It's cloudy tonight and an occasional patter of rain drops comes on our roof. It sounds sort of lonely. . . .

Won't it be nice when the war is all over and we can get back to normal again? Or do you think things will ever be the same again. Well, perhaps not just the same but almost.

I wonder what we'll all do then. Every once in a while I come up with a start and remember that I really know nothing about housekeeping and really so little about anything that you'll be up against an awful proposition. I wonder how much you'll mind. - - - - - (indicating wonder)

* *Mary Blampied*

** *Art Hall, the "embryo lawyer" mentioned earlier in Letter #29*

Every one was out looking and listening—great sport this watching for air raids—but we went on doing just as we had planned. We went past one bunch of tents past a big bunch of men, crossed the road & just as we crossed the corner something went z-z-z about 5 or 10 feet behind us. It sounded like a dud shell from the anti air craft guns. We ducked and ran. Such a loud yell and laughing!!!

Cook house.

Reinforcements Arrive

Memoir of Reserve Nurse May Connard

When I arrived in Chicago [around May 1, 1917], I found that the nurses with whom I had lived were preparing to go with a Base Hospital Unit to France in a couple of weeks. They were giving up their apartment, and very excited about leaving. A Reserve Unit was being formed, which would leave in the fall if needed, so I quickly signed up for that. The Reserve Unit was called the first of August 1917 and sent to New York to await orders. We stayed in an unused hospital ward on Ellis Island, where the immigrants had previously been quart[er]ed. Great secrecy was maintained as to when we would sail. Each morning we reported and then were free until the next day. There was nothing to do except go to New York on the ferry, which became rather expensive. I had to go before the British consul, to answer many questions about my service in Germany. He must have believed that I was no spy and knew nothing of value to either England or Germany, for I was allowed to go.

We remained at Ellis Island two months, until our orders came to sail on the Carmania a sister ship of the Lusitania which had been sunk. Nurses and officers had the first and second class cabins. There were more than 1000 troops aboard. After two weeks dodging mines we docked at Liverpool, and took a train through England to Folkstone. We changed trains at London but saw nothing of the city. At Folkstone we took a ferry across the choppy north sea to Boulogne, France. After a night there, British lorries came to take us to our Base hospital near Etaples, France. It was about 15 kilometers away.

The hospital had been set up by the British, but was now staffed by U. S. personnel. However, equipment and food were supplied by the British. There was a British major on the staff, as liaison officer. Quarters for living and offices and three huts for patients were wooden structures. The rest of the patients were in large tents, in all about 1000 beds. This was one of the first base hospitals back of the field hospitals. Patients were usually brought in at nights, driven in convoys without lights. They had received only first aid emergency treatment, so we were very busy after a convoy arrived. As soon as a patient was able to travel he was sent back to England. There were many amputations, compound fractures and other severe injuries…

Subsequently, the authorities erected better buildings, quonset-like huts with sandbags around exterior

Letter #36

Laura to Erle
Sunny France
Oct 7, 1917

Dearest Buddy:

Today has been such a day—so maddeningly busy inside and so wildly windy and rainy on the outside that I scarcely knew where I was half the time. But this evening the rain finally ceased so after tea I sallied forth alone to take a little walk. The clouds are always beautiful here. They were gray and scurrying but a streak of yellow along the horizon lightened their moodiness. It was just dusk when I started and quite dark when I returned. Miss U is away ill so that accounts for my going alone. If she fails to disuade (?) me from my purpose she drops whatever she has and goes along. So you see I'm well protected. . . .

Last night a bunch of us went to a funny little hotel in the village for supper or rather dinner. We went in thru the café and up some funny twisty stairs to a little room just large enough for our table. We had a red checked table cloth and lots of good things to eat. It was lots of fun and we had a nice time.

It's getting terribly cold here. We almost freeze up when we sit still long enough to rest a minute. I hope it warms up soon. We've really had a lot of fine weather besides lots of rain.

I don't know just how many hospital units there are here now but when we came there were six. One other is our neighbor and we met the nurses from a second in New York. The nurses of all three units are of a type—practical, self-reliant, capable, professional. I have seen many who like having lots of fun and lots of men friends, a few who incline to be frivolous but the latter are the exception. Is that sufficient answer to the tale told you about nurses returning to America. We had heard this same tale before—I presume started by some vicious poison-minded person. It is a shame for some people always believe anything they are told. To us here it is the height of the ridiculous.

A few nights ago we had a party with coffee as the main part of the lunch. It was delicious. I can't tell you how good it tasted to me and to the other girls. If I can get some fuel, I'll have some more tonight. It was sweet of you to send it. . . .

I told you in my last that our reinforcements have arrived and we're nearly working them to death & cruel & inhuman treatment!!!

I've had a letter and a package of woolens from home this week. If you thot I was hinting for you to send some to me you'll see that I had applied elsewhere for aid. I do have a perfectly good family, you see, who are only too anxious to send me anything I'll mention.

Later

The supper bell rang just then and I went over to eat. There was a bunch of American mail. I had two S.E.P.'s and a letter from Miss Stodola. I'll enclose the letter just for a sample.

I wrote her one short note soon after arriving here. You know she had had me out to dinner and had sent me two boxes of fudge before I left Chicago. I've had a letter about every week from her whether my other mail comes or not. I hope her affection dies down soon. It irritates me. But then these several thousand miles between us is considerable barrier.

Erle to his mother in Indianapolis

CHICAGO

October 8, 1917

Dear Family,

Have you just about decided that I've been spending my time in jail the last few weeks, without privilege of communicating with the outside world? Well, it hasn't been as bad as that, quite, but my time has been used up so completely that in order to write this missive I have played hookey from drill tonight.

You see, the members of our unit now take two evenings of drill each week, two evenings of surgical aid instruction and two evenings of conversational French. They have Sunday evening for wild carousal. In short, outside office hours nowadays, I have practically no time for my own devices. So do you forgive my negligence?

Need I express my very great hope that you are by this time nicely settled in the new home and that things are running smoothly? I have been thinking so much about that important change that I'd take a train and go down at once if I felt I could spend the money and if I had the time. If we don't go into active service pretty shortly, I will go down to have a day or two with you—if I have to walk back to Chicago. I'm more firmly convinced than ever that the establishment of a new home is the very best thing possible for all of us—even though we should have to pinch for a while. We must continue to forget the word "if" and force ourselves to have confidence in our ability to pull through and faith that the future will bring happiness and prosperity. So much depends on the kind of thoughts we hold in mind, all of us; doubtful thoughts breed doubt and positive thoughts produce achievement.

I believe I sent you copies of the correspondence that passed between Red Cross headquarters and me regarding my doing publicity work when I reached Europe. I forwarded copies to Major Doane at Fort Harrison. He sent me a cordial, encouraging rep1y—copy enclosed.

Friday night, our unit dissipated a bit by having a dance at the Chicago Yacht Club. All the men were in uniform, though the affair was informal, and everyone enjoyed himself, I should say. I went stag, but there were extra girls on hand for the stags. The officers and their wives came, too, and we privates had the Opportunity to meet them all as social equals. Of course, the weather man conspired against us—it rained torrents and hailed, but we used ambulances as cabs from the club to the street car lines.

I'm sure I told you I've moved. Now I room at one place and have my meals two doors away. The arrangement is more satisfactory than that at the Alwards. I have a comfortable room, the meals are very good and my fellow boarders are pleasant people who mind their own business.

Two or three weeks ago the Western Newspaper Union bought out its competitor, the American Press Association, and effected a consolidation. Since then all of us have been very busy readjusting things. Mr. Patterson tells me the company is going to invade the metropolitan field and issue big syndicate features for leading newspapers like those in New York, Chicago, Philadelphia, Boston, etc. We have signed up with Col. House, president Wilson's mysterious friend, for a series of articles—the first stuff the press has obtained from him—and we are negotiating with ex-president Taft for a series of articles.

Oh say, Hettie, have you done any work for the editorial department? I've been wanting so much to write you at length in answer to your letter regarding the matter. All you need is confidence-honestly. You write an excellent letter. Your intelligence is much above the average in newspaper offices. Once you start, and forget your self consciousness and timidity, you will sail right along. Jump in, Sis!

And Betsy, what would we do without you to be the man in the family—bless your heart!

I get dead homesick for you all very often these days. . . .

Letter #37

Laura to Erle
France
Oct. 13, 1917

Dear Buddy: —

Of course you don't know that I have started another letter to you about three times this week and didn't get them finished. One I tore up because it wasn't cheerful and I thot again & didn't send it—you see we've been so busy and things don't always go just right even in such a nice ward as ours. The Medical Officer was cross and for awhile I was considering asking to be put on night duty which is the same as solitary confinement or at least as bad.

We have had rain, rain and more rain all week and we are beginning to wonder if it's ever going to be warm again. Off duty time is spent hovering over a small oil stove trying to be comfortable.

Tonight we had coffee again. Honey, that coffee has been a great pleasure to me —and to my friends. I can't think of anything that would be as much real joy to me that I may have here.

Last night we had a dance again. As dances they don't amount to much but we do need something to help us relax a bit. I think they are very good for us. . . .

Sunday [14th]

I told you before about Miss Urch being ill did I not? She isn't doing at all well but she is in a delightful place. It is a villa built by an American millionairess who married a Servian Count and lived with him three weeks, left him and returned home to mother. The count is here off and on but has turned over the villa for the use of sick nurses, V.A.D.'s and so on. Miss U. has the mother-in-law's room. She did have the Count's but was moved from there.

This is a great place. Some of the older "girls" who haven't had beaus for years are stepping out so gaily and giddily for their years. I think sometimes that three fourths of them are man crazy but suppose it's really but one way of relaxing.

We are so cold here now. I wonder how we'll ever keep warm when real winter comes. The cold is so damp and penetrating. Sometime when I get to be rich I'm going to go to a tropical climate and camp there indefinitely just to get warm.

You remember how it used to be when a cold snap came in May after all the stoves were down. Well that's our chronic state now. . . .

You know I don't think this system of organizing a bunch of young professional men and millionaires' sons into a group of hospital orderlies is altogether successful. But it certainly shows what kind of stuff the boys are made of. As a group they do fairly well but nothing to brag about. . . .

You know I don't think this system of organizing a bunch of young professional men and millionaires' sons into a group of hospital orderlies is altogether successful. But it certainly shows what kind of stuff the boys are made of. As a group they do fairly well but nothing to brag about. . . .

Pvt. Baker's memoir:

October 15

Our reinforcements of nine M.D.'s, thirty-three nurses and forty-seven men just arrived after being on the way since September 18. I had met a couple of these men before. There was quite a mixture of personnel. Several were from the Harvester Company at Chicago.

The greatest drive of the war opened and is still going on. We have received several hundred wounded boys already. All have the same solemn expression on their faces. Each carries his "dorothy" (little cloth sack) containing his personal possessions such as toothbrush, towel, soap razor, badges, etc., and which represents all that is left of the things with which he started out.

When a man enters a hospital, all his clothes, with rare exceptions, are taken and fumigated and he never sees these clothes again. He is given others until he gets either to Blighty (England) or back to his battalion's base. I have seen many men wearing Fritz belts and bought one from a man who had just come down from the front. Many bring revolvers and a few bring field glasses taken off German officers and artillery men.

It is pitiful to see all the wounded that are able to walk, lined up for their meals. Each has an injury—an arm bandaged, or a foot or neck or head, or some other of a hundred different kinds of wounds. They stand in line for two hours before all can get in to eat and never a grumble. Patience seems to be their outstanding virtue. I can not imagine the same about the Americans. The British Tommy has to be led, the Yank would rather do his own leading. In the ward, the Tommies are ever patient, orderly and very good helpers. Of course they know they have to be that way anyway as the nurses and orderlies are over them.

It is always a pleasure for me to look at the photos the poor fellows always carry with them of their mothers, wives, children, etc. The Tommy is always ready to tell you about his wife and kiddies and show you the different photos of the children and how they have grown since he last saw them, probably a year and a half ago. Some have shown me photos of their new babes that are probably six to ten months old and whom they have never seen, but the wives keep them well posted. It seems hard to think this true. Here they are less than fifty miles from home and cannot get back unless badly wounded or something equally serious has happened. "War is Hell" surely as Sherman hath well said.

October 16

We were awakened by heavy firing about 12:30 this morning. It was very dark. There was a Boche machine up there somewhere and the shells from our guns whistled over our tents for about twenty minutes. We could not see a thing. I was laying with my tent mates flat on the floor of our tent face down so that if a bomb hit nearby I would be missed by the flying shrapnel. We expected a bomb or a dud any minute. Nothing was harmed that we heard of.

Soldiers and their plunder.

Letter #38

Laura to Erle
France
Oct 16, 1917.

Dear Buddy: —

. . . We received real beds yesterday and I surely was delighted. They are small and black and have springs and mattresses and are really comfortable.

Elizabeth and I moved ours together so our blankets would go over both and we were warm as toast last night for the first time in two weeks.

You know it isn't really so cold yet but the dampness is so penetrating.

It has rained almost a week now. Everything feels so musty and damp.

We had a death in our ward today. It was not so sad as some others but it certainly got me just the same. I wish that such things didn't bother me so. . . .

Did I tell you that we have a little oil stove in our room now? It helps the atmosphere quite a lot. . . .

> *Note: Russian Revolution, October 17, 1917—Imperial Russian government overthrown by Socialist/Communist rebellion. In a few months, this will allow the Germans to shift forces from the Eastern Front to the Western Front, resulting in dangerous setbacks for the Allies.*

Diary of Dr. Neal:

Oct 18—Zeppelins in a body over our area near midnight—made a raid on London.

Pvt. Baker's memoir:

October 19

About 12:15 a.m. when all was quiet and everyone was asleep the air raid warning of three blasts of the whistle sounded. A few minutes later, boom! boom! went the guns ten miles north of us. It was very dark. The boys beat it for the trenches nearby. Every few minutes I could see the lights from our guns. The enemy plane came straight over our heads and all the batteries around us opened up on him. The shells from the nearest gun would leave the gun and go "sizzzzih" and then "puh!" as the shells would burst. There was hurrying and scurrying around camp. I saw a half dozen men beat it to the hills as we had done the first time. We three in our tent rolled off our bunks and onto the floor and lay there till all was over. As luck was in our favor he never dropped any bombs near our camp, but left them all at Boulogne before he reached us. I was sure shaking all over from being awakened so suddenly, by the chilly night air and the thoughts of the duds falling all around.

The Sanitary Squad challenged the Q.M. Department (which I organized) to a football game and after three practices, we played. The Colonel sanctioned the match and was out watching the game... We had every imaginable kind of garb on. It looked like a real hard-times party at the start... As far as known, this was the first American football game ever played in Europe, or the Continent.

Letter #39

Laura to Erle
France
Oct 19, 1917

Birthday greetings—Congratulations etc. I celebrated it by a half day. I walked to the sea, came back by the village, spent some filthy lucre for grapes for my boys, to the mess for tea, walked up the hill in the dusk and spent the evening fixing up some things for our room. Some celebration isn't it? I wonder if you did as well. . . .

Did you ever have chill blains? They are our latest malady—among the sisters. I have a few on my feet and many on my fingers—or else it's something due to the diet here. Anyway we have something that isn't comfortable.

Bertha told me that they had received their first mutilated letter ***[censored]***. Probably the one I sent you at the same time was likewise treated.

I went to town again today. I purchased one shampoo, one small red rug, one towel rack, one small cameo and a piece of lingerie. I'm picking up some little things for the family as I find them. The shops are small and have very little to sell. But I enjoy going about anyway except when I have something definite to buy. It is next to impossible to find just what we want in anything.

We've had a lot of rain this week but tonight is beautiful—clear moonlight. The air is not like our Autumn air but cold and crisp and clear like February weather. . . .

Pvt. Baker's memoir:

Saturday, October 20

The Sanitary Squad challenged the Q.M. Department (which I organized) to a football game and after three practices, we played. The Colonel sanctioned the match and was out watching the game. After we lined up before him, he made a little announcement to us. We had every imaginable kind of garb on. It looked like a real hard-times party at the start. I had a sweater with pacts in shoulder and elbows, long slack pants, a knit cap with socks in it for headgear and wool socks. Harold Hanish of the University of Chicago (1916-17) was referee and Fred Norman of Northwestern University was umpire. Everything went off fine and no one was hurt. We won, the score being 12-6 and the spectators seemed to enjoy it. The British wondered why a lot of us were not killed and why more were not hurt since the game looks so rough to them. Some of our men brought stretchers out on the field as a joke. The English all thought this a regular part of the game.

As far as known, this was the first American football game ever played in Europe, or the Continent. I received a strained ankle and Charley Horse on my thigh. I played right halfback.

The following week was all rain with patches of sunshine. The moon was getting larger and one night while we were helping an evacuation (carrying wounded men to ambulances to be sent to England), the moon was out while it was raining hard. There was a rainbow clear across the sky. It was very peculiar. I had never seen one before. We never see anything like this in America. In France it does not lightning and I am real homesick for some lightning and loud thunder.

The Cross-Atlantic Mail Lag

Letter #40

Laura to Erle
Thursday [Friday] 26th

Dear Buddy:-

. . . I woke up this morning and decided that if many more weeks went by without a message of any sort I'd write your Aunt Sadie to find out if anything had happened to you—an automobile accident, street car collision or what. I've just been thinking of what all might happen and I'd never know till many weeks or perhaps months afterward. Worry is a fine thing, is it not? So I'm not going to worry. Perhaps I'll have a letter within a few days.

We've been having wind, rain and work here and but little else. It has been so chilly and damp that we've kept in our rooms most of the time reading or writing or talking or drinking coffee.

My how I wish you were near enough to take some nice long walks with me. The moonlight is beautiful tonight between the clouds. Perhaps it will be raining in another hour but just now the moon is brilliant.

We are none of us very flattered at the prospect of a charity Xmas dinner. We are all perfectly able and willing to buy our own dinners and we don't much like the idea of that publicity-crazy dame giving a benefit dance for us. I think she's a bit too fond of the lime light.

We are going to have a Halloween party in our mess. I'd like very much to have you as my guest. Do you think you can manage to come? Every-one is to be masked and the chief amusement will be dancing.

Yes, I know it's war time but even so the human animal must be amused. I wish you could have seen the line up outside the Y.M.C.A. Hut waiting for the doors to open for a concert. And the hut is so small there's never room for half the people who want to attend. I had never realized before how people crave entertainment. It's too bad there isn't some sort of arrangement made for movie shows in all the Camps. Some of the men with superfluous millions should be able to arrange it.

[This is Laura's second reference to "millionaires" and gives an indication of the societal class distinctions between the nurses and their subordinates, the University Men, who almost by definition in those distant years were privileged members of urban society, in contrast to the nurses and their small-town backgrounds.]

Sunday the 28th

I've just come from a nice walk in the moonlight. We walked to the next village to buy a little flashlight for my nice young Canadian who has been doing "light duty" in Ward S. He is such a reliable boy and I like him and I do hate to have him go back up the line. But as perhaps you know there's a war on. . . .

Monday Evening

Another perfectly gorgeous moonlight night. The line of the hills is distinct for a long way, the stars are brilliant and twinkle and the moon is most beautiful of all. We have just returned from a lecture by a Scotchman on "America in Arms." It was very good and interesting but the walk home was better yet. Last night's mail brot me two long delayed parcels from home and a roll of newspapers of Oct 6th from you. So I know that altho I have heard nothing from you since Sept. 20 that something over three weeks ago you were still alive.

Honestly, I was so disappointed to receive no letters last night nor today but they'll surely come soon. I've been thinking that for three weeks now.

I'm afraid most of this letter is complaints about the mail service. I haven't much else to write. Tomorrow is inspection day and that will be awful I'm sure. It always is. . . .

Pvt. Baker's memoir:

November 1

At 1:30 a.m., the air raid warning blew again and immediately we heard the guns booming fifteen or sixteen miles away, getting nearer all the time as Fritz traveled our way. He came to within a few miles of us and then turned out to sea and back to Boulogne where he stayed dropping bombs. The "air" guns boomed for 1-1/2 hours. Poor Boulogne!

The English are pounding the line daily in the hopes of breaking it. We sure see the effects of it here at the hospital. Just seeing the wounded and sick men all the time grows so monotonous that we have to do something else like athletics, to keep from going cuckoo.

Diary of Dr. Neal:

Nov 2—first Americans killed in action during a German trench raid on our sector.

Letter #41

Laura to Erle
France
Nov. 3, 1917.

Dear Buddy: —

At last the mail has come. Did you know that I was twenty seven days without a scratch from you? But it's nice to get it in a lump, too. Thursday I had two letters dated Sept 25 and 30th. Friday I had two more dated Oct. 7th and 12th and today another of Oct 11th. Also three rolls of papers have come. So you see I've ceased to feel neglected.

And I've had six letters from home, one from Helen K and several from Miss S. and a box of candy from Miss S. which I have not yet opened! . . .

Theaters and operas, etc. seem many leagues away from us here. But we're too busy anyway. We've had no let up for a month or more. The past week has been nerve racking for me and I've so longed to give up the ward and just pitch into the work and let some one else do the bossing. I hate being in charge. It wouldn't be so bad either if one could be only in charge but you see "in charge" is just extra responsibility and work besides the routine of dressings, bed-making and care of sick patients. . . .

And how do you like French? I've given up trying to learn. I'm not brilliant in languages at best and since we've been so busy I simply couldn't take time to study nor to go to class. And, besides we need it so little here. Almost every one is English or speaks English and there's always an abundance of interpreters if one is needed.

I was going to write you about the nice rainy visit I had with Miss Urch last Tuesday. It was cold and rainy but we drove over anyway. The drive was lovely and the visit or rather call delightful. I wanted to go alone but Miss Cleveland decided to go also and I was glad to have her along most of the time.

I'm so anxious to have Miss Urch back with us but I'm afraid it will be another month anyway. [***The nature of Miss Urch's illness, requiring a long convalescence, was never disclosed.***] . . .

We had a very nice Halloween party but I staid only a short time and didn't mask as I had worked all that day without any time off and was too tired to care.

Now, don't think I'm working too hard. I'm not. And besides that it's war time and we must all do our bit. . .

Diary of Dr. Neal:

Nov 3—Italians retreating en masse. Huns report taking 200,000 prisoners and 1800 "big" guns.

Letter #42

Laura to Erle
France
Nov. 8 1917.

Dearest Man: —

I'm having a half day and I'm just loafing and resting. We've been so busy, busy, busy. But I believe I've groused a lot about that before so I'll not mention it again.

This afternoon Emma and I went to the village to buy parafin oil for our stoves. I had a three litre can and decided to purchase another five litre one to store away. Can you imagine the picture I made coming home carrying eight litres of oil in the two tin cans?

On the way to the village we met such a nice old Scotch captain and he told us to lay up a supply of fuel by fair means or foul because there'd come a time when the railroad will be tied up and we'll be unable to get fuel for a week at a time.

Having already had a number of chill-blains on my hands and feet I didn't care to repeat the process. I'm going to store bottles of oil and boxes of coal if I can get it under my bed. We haven't begun using our little coal stove yet. I'm not a very good fire builder and the oil is so much cleaner. We'll probably have to use both to keep warm when the real winter comes. . . .

Diary of Dr. Neal:

Nov 7—Fourteen U.S. Congressmen visit our unit.

Letter #43

Laura to Erle
France
Nov. 10 1917.

Dearest Buddy: —

Absolutely nothing has occurred since my last paper visit with you so don't be expecting a newsy letter. There is no news.

Of course we've heard rumors of the Russian troubles and read of the Italian retreat and surrender and it takes considerable starch out of us.

Nov. 11, 1917.

I decided that was a poor start and anyway I had nothing to say so now I'll try again.

I wish you could see your La-la tonight. She is so comfy and clean and warm. She's just had a nice hot bath and is now all curled up in our one chair beside our hearth stove which is a "double Beatrice" oil stove. She hasn't turned on the light so there is just a pleasant glow. And she'd love to talk if she just had an inspiration as to what to say.

Everything went off beautifully today. We were thru our dressings and bed making by noon so everything went off beautifully of course. I do love my ward when it's nice and I hate to think of turning it over to anyone else as I'll have to be doing one of these fine days.

When I came home at five o'clock Bess had a fire in her coal stove and we made coffee and had crackers (known here as biscuits) and fruit cake that Elizabeth had gotten from home. . . .

We do so enjoy the coffee you sent me. But if you don't hurry up I'm afraid you won't get here for a cup of it. We've started on the second tin.

Today I brot home from the ward two pairs of nice long knitted wool socks. They were meant for wounded soldiers going to operation but I decided that I needed them also. They are so warm and soft. I'm going to wear them in my room with my woolen bathrobe.

Please don't worry about me and my comfort. I'm very comfortable and there's nothing at all to worry about.

This week I received three large rolls of newspapers. I tried to read them all at once last night but had to give up and go to bed. It was certainly thotful of you to put in the novels and the magazines (Red Cross & Fine Arts). Many thanks for the same. . . .

On the way to the village we met such a nice old Scotch captain and he told us to lay up a supply of fuel by fair means or foul because there'd come a time when the railroad will be tied up and we'll be unable to get fuel for a week at a time... Having already had a number of chill-blains on my hands and feet I didn't care to repeat the process

Letter #44

Laura to Erle
France Nov. 12

Dear Man: —

I've just finished reading "Christine". Have you had an opportunity to read it? It's very interesting. Conny was here while I was reading it aloud to Elizabeth and she was very much interested and said it certainly was a faithful representation of the German people as she knew them. . . .

Elizabeth and I both had half days today. We walked to the village to buy oil again. We bought it from a funny old Frenchman who had it stored in a tank in the garden. We went thru the house to the rear and it was so funny. There was a huge fireplace of blue tiles and a stove stood in the fireplace. That seems to be the usual way here. The houses very old but the people are modern enough to use stoves. The peasants dress just like pictures, too. You'd enjoy seeing them all, I'm sure.

After we had bought the oil and paid for it they asked if we'd like flowers. Of course we would as the old man limped back to the garden again to cut us a nice big bunch of Chrysanthemums.

You can see that it isn't so cold here yet or the flowers would be frozen. I don't believe we've had a frost yet but the air is so damp that we chill a great deal. But taken all in all we're very comfortable especially here in our own little room.

The marquees [***tented wards for the patients***] aren't nearly so cold as you'd think. Unless the wind is too strong we can get a good fire and it changes the whole atmosphere of the ward.

Miss Urch is still on the sick list. She is going on a three week's leave of absence to southern France before she comes back to duty. I'd love to go along.

I've asked for my vacation the last week of February. Please look for me in gay Paree at that time. I do want to go to southern France a bit, too, if possible but I'm not at all sure how far they'll let us go. Six of us may go at one time so Emma, Bess, Frances, Ella, Elizabeth & I are going together.

How I do run along! I did just start out to tell you about how well I liked Christine. I don't owe you a letter at all. I just mailed you one yesterday or was it this morning? But I'll keep this over for a day or so and write some when I have news or else when the spirit moves me to talk again. Good night. There's Taps. I'm afraid we don't pay much attention to regulations here. We stay up as long as we please and keep our lights on but of course we are privileged characters. I'm usually too sleepy to stay up late. I think it's the much fresh air that makes me so sleepy. . . .

Thursday [Nov. 15]

I'm getting very lazy these long evenings. We seldom walk farther than to the village because it gets dark so early. We've even been neglecting supper & making coffee over our stove.

This evening I came home a little early from work as we weren't at all busy in the ward—moved the wash stand off the tiny coal stove and had a general house cleaning. Hereafter we are going to use the coal stove if we can make it burn. We've decided to store up the parafin for emergencies and not turn our boudoir into a coal yard. . . .

Photo of Laura with two other nurses, beside one of the huts.

Pvt. Baker's memoir:

Sunday, November 18

In the morning, while at work in the sorting tent, we saw a British plane coming very low and we watched it dip down just as it reached our camp. Then it arose and as it got to the main entrance of the hospital ground one-quarter mile farther, we saw it bank very sharply and turn in a circle. When about 300 feet up in the air, he pointed his nose straight down, his engine still running. He turned in spiral form downward and just as he hit the earth, he attempted to spread out but his control was lost and he crashed into the ground. His body was crushed so as to make it very hard to extricate from the plane and after he was taken out he died on the stretcher on the way to the officers hospital. I was there about two minutes after he hit. There were patients who sprinted over there and forgot they were sick or wounded, just carried away with excitement. A crowd of about three or four thousand gathered in a few minutes. The aviator had just come across the Channel from London and was a Canadian Ace on his way to the front. It was a terrible bloody sight to see him while they were taking him out. His seat frame and foot board were smashed, as was the plane. The motor had dug up two feet of ground.

In the afternoon, Ed Kahout, Shorty Gilmore and I lit out on foot for Hardelot-Plage, an old seaport, fishing village and wine center…

• • •

We left Hardelot-Plage after going to the center of the town and listening to a band concert at which we three "Yanks" again became the central attraction for all the wounded soldiers and even the band leader. From the way they looked us over, it was the first time these Allied soldiers had seen any "Sammies" or Yanks. We made our way back along pitch black roads in about one and a half hours. Captain Nadler of our unit came back with us. He is a regular walker. I nearly ran all the way to keep up with him. We ate in an officers restaurant and got away with it O.K. Captain Nadler is the best friend we have in the outfit.

In the morning, while at work in the sorting tent, we saw a British plane coming very low and we watched it dip down just as it reached our camp… When about 300 feet up in the air, he pointed his nose straight down, his engine still running. He turned in spiral form downward and just as he hit the earth, he attempted to spread out but his control was lost and he crashed into the ground. His body was crushed so as to make it very hard to extricate from the plane and after he was taken out he died on the stretcher on the way to the officers hospital. I was there about two minutes after he hit.

From The Chicago Daily News, sometime in December 1917

CHICAGO CAMP SEES CANADIAN'S SACRIFICE

Steve Ketchel's Former Opponent Proves Himself a True Hero

BY NORMAN KLEIN

[Special Correspondent of The Daily News.]

Chicago Base Hospital in France, Nov. 18. -- Who says there are no modern heroes?

Sergt. Steve Ketchel, Chicago light-weight prize fighter, who has charge of the carpenter shop in this camp, took part in some ring exhibitions staged the other day by soldiers in a little French village not far from here.

When Ketchel was climbing over the ropes, after winning a five round bout amid the shouts and cheers of hundreds of Tommies, a young man in a ringside seat jumped up and shouted to attract his attention.

The Chicagoan turned his head and saw a handsome six foot chap in the uniform of a Canadian captain. He wore one of those khaki caps on one side of his head—it looked like a Jock's glengarry—and Ketchel knew the Canadian was a pilot in the British flying corps.

"Hello, Ketchel!" called the young Canadian captain. "Don't you know me?" Ketchel was off the platform by this time. He slip-slapped in his ring shoes over to the captain. And when he had waded through the throng of soldiers and the fog of tobacco smoke Ketchel's face lit up with recognition. Sure he knew the man! Had not he battled against him for ten rounds in Toronto one winter while on a tour of Canada?

And so the two men shook hands and laughed and talked of old times. They never expected to meet one another over here in France, on the western front. Several times they met after that, and the other day the Canadian airman told Sergt. Ketchel to be ready, for he would call for him in a British scout airplane for a trip over the lines.

It was a Sunday morning. The weather was severe and sunny and the sky was clear. We stood watching an airplane come droning with the noise of a saw-mill high in the air toward the camp. The motor was throbbing and roaring in a business-like fashion.

When we first sighted the "bus" it was but a tiny dragonfly. I might say here that sometimes in a day a dozen airplanes pass over the camp, but if we get a chance we always poke our heads out of the tents to watch them; they fascinate us somehow with the sureness of their flight.

Suddenly the airplane tilted downward. It wavered a moment on the brink of some air precipice, and then it plunged down, down. Far miles and miles the airplane seemed to dive through the sky, down toward the sand dunes over by the sea. It was not really miles the airplane fell, of course, but it seemed so. And our hearts gave a flop and stuck in our throats, and we could not breathe somehow.

The pilot was in trouble. It was plain to see that. Had the engine failed? Or were some essential rods broken? Or had the pilot fainted from "air sickness?" We never knew. Nobody ever will.

He righted the machine in that dizzy, roaring plunge, when only 500 feet from the ground. The machine "banked" and shot straight ahead. Then up it scooped

It was directly over our heads that the airman stopped the downward dive in some miraculous way. It was a quarter of a mile north, over another hospital, that he succeeded in turning the course of the machine and sliding up into the sky again.

And then I had a hunch and ran with all my might in the direction of the troubled airplane. I elbowed wounded soldiers in blue and men in khaki and kept my eyes on the now wavering thing up there in the sky.

Without more ado the biplane stopped going up; it began to whirl around and around, chasing its tail like a crazy dog. Then it dropped like a rock into a pool of water. Down and down it fell swiftly and disappeared behind the horizon of hospital tents.

There was the noise of tearing and crashing; that was all.

I was the first American on the scene. With its propeller nose buried in a potato patch, canvas wings broken and crushed like so much cardboard, the airplane lay there; mere rubbish.

The aviator we pulled out of the wreck of engine, petrol tank and wire and wood supports. He was caught pitifully in the trap of the pilot's seat. I had a glimpse of his face, handsome and firm, and a bit oldish, because men grow old swiftly nowadays in this part of Europe. Not until afterward did I realize that I was covered with his blood. He was dead. They covered him with the Union Jack.

And then I met Steve Ketchel. One question was enough. The dead pilot was his friend, the Canadian, who had promised to take him for a ride. He had come to call for the Chicago boxer when the airplane fell.

In order to keep from plunging on some of the hospital wards in this camp and killing nurses and patients within, it was said that this young Canadian birdman had flung his troubled airplane at the only empty plot or ground, the potato patch. And he knew it was suicide!

Weariness, Wounded and Infections

Letter #45

Laura to Erle

This letter, in a tan British Expeditionary Force officer's envelope with dark green printing, carrying a One Penny Great Britain stamp, and addressed to John Erle Davis, Western Newspaper Union, Chicago, was typical. It was sealed with a gummed label by censor #3531 and marked on the reverse in Erle's script "Rec'd Dec 19—17."

France

Nov. 17, 1917

My dear Buddy: —

I'm off duty today with an infected finger—very romantic, almost as much so as measles—the first one of our family to lay off for a minute. I'm ashamed of myself and will try to do better next time. I've been soaking it in hot water or else wearing a hot dressing on it all day. I hope by tomorrow it can be opened.

Nov. 20

Of course you don't know that I haven't mailed you a letter for a week but I'll tell you so you won't be looking vainly for the one that I didn't write. You see it was like this. I have soaked and hot dressinged steadily for three days and yesterday had my finger opened up and the pus let out. So now I'm all right except that I must still soak and wet dressing for a day or so.

I've had lots of time to read and think and dream and I've done lots of all three. Last night we had such a delightfully hot and bright fire in our little coal stove that we turned out the light and just sat and watched the coals and the flicker of shadows on the walls. I don't know what Miss Cleveland thot about when she came home but you can guess where I was in spirit. It took me back, too, to days or rather nights at home when Edna & Edith [North Vernon high school friends] used to stay all night with me. We'd build the fire up in the parlor stove and open the door and sit in the firelight and dream and wonder about the future and about so many things of which we knew nothing—philosophy, theology, love, men, women, life and many lesser subjects. We were a combination, the three of us and yet we had a common love for the beautiful and a vague something that we could neither reach nor understand. We'd sit up almost all night and then go to bed and talk till daylight.

We seized every opportunity for such a session but at Tech's and Weedman's [her friends' family names] there were younger children to be entertained or amused or to creep up to hear us talk. At my home we had the parlor and bedroom all to ourselves. I sometimes think that my mother understood us much better than we did ourselves even tho she seemed not to.

It isn't a great while now till Christmas. I thought back over different Christmases I've spent and strangely enough none are at all distinct till I came to Chicago. Before that all is a blur. I wonder, did I just begin to live then?

Last year I spent Christmas Eve with the Kelleys . . . In 1915 you and I went to St. Mary's to midnight mass. I shall never forget the snow storm and the Christmas carols we heard and then the mass and the choir boys and flickering candles and the trip home afterwards. In 1914 Uncle Wright & Edna [***her first cousin***] were at N.V. for a Christmas with the children. I went home for the reunion and we had such a happy time. It was my uncle's last visit to us. Before the next year he had passed on also. (He had lost his wife just a couple of months before he came to us). [***Wright Huckleberry had been the chief of police and fire, and a coal-and-ice merchant in Cadillac, a lumber town in northern Michigan.***]

1913 I was home for the whole holidays, the first time since I had left home.

1912 I was on night duty at Children's Hospital. (Happy memory of dolls and toys and pink cheesecloth stockings on every crib). 1911 night duty in old ward 26 a men's medical—sponges and tubings and packs by candle light and voices in the corridor singing "Holy Night". 1910 Edna [***Tech, Laura's hometown schoolmate at Illinois Training School for Nurses***] had typhoid fever. I was wakened very early in the morning to come in to help her open her packages and to have breakfast with her as a special treat.—Just a series of memory pictures perhaps not interesting to you, but I tell about them anyway.

This Christmas will be different from all the rest and I hope next year will be still different. I can plan a delightful one for a year hence. I hope there will be an open fire and not too many people. . . .

A couple of days ago four of us (all girls) went for a walk out across the hills to a little village for supper. We lost our way going and went across plowed fields before we finally reached the road. It was a chilly gray day but we didn't mind either that or the mud. We are accustomed to both now. The village was delightful except for its filth. The low houses and tiled roofs along the lane—you couldn't call it a street. There were scraggly trees and low walls and high walls and the inevitable barnyards.

Then after an excellent supper of brown bread, butter, cold roast pork, stewed quinces, salad, browned potatoes and pear marmalade we started home. It was pitch dark and I don't know where we'd have landed without a flash light. When we got out of the village on to the open road it was easy traveling. You'd have enjoyed it so much—the hard road over the hills, the tiny scraggly row of trees on one side, the occasional searchlights or perhaps just a glimmer from some town or camp.

You and I could have such good times here. We enjoy so much the same things. When I'm a millionairess I'm going to tour France in an automobile. Will you join the party?

This afternoon I'm going over to see Miss Urch. She is still at the Villa Tina and will probably have to take a leave of absence before she'll be able to come back to us. I do miss her—and yet we spend very little time together. But after all it isn't the time one spends with one's friends that counts.

Did I tell you that I'm going or hope to go to Paris the latter part of February? Perhaps I can manage to go to some other places as well but I'm planning only on Paris.

I'd like so much to take you with me this afternoon to see Miss Urch. The road out there is mostly thru the woods—not big trees—but brush and little pines and whether one walks or rides it is very enjoyable. . . .

Evening

Miss U. goes to the Riviera soon for three weeks. I'd love to go along but of course I can't.

Tonight we had coffee again in our room. Honestly, I can't begin to tell you the pleasure that we've had from your gift. Tea is so sloppy. And since we have to have a fire to keep us warm it is so nice to put on the pan (we don't have coffee pots here) and make a few cups of the brew.

Diary of Dr. Neal:

Nov 20—400 British tanks followed by cavalry take 10,000 prisoners and 100 "big" guns in advance around Cambrai—penetrated 6 miles to or near the Hindenberg line.

The Receiving Tent, where the wounded and ill, fresh off convoys, were assessed for injury or illness, perhaps had intermediate trauma treatment and were assigned to correct ward for treatment. Patients treated by BH 12 from June 1917 to Feb. 1919 numbered about 60,000, about half in each category (sick or wounded) and some in both.

It seems so funny to get your letters and read your comment on my mood of two months ago. I'm funny to myself when the moods change so I must be more so to you. . . .

Letter #46

Laura to Erle, Nov. 21, entirely personal contents unrelated to war experiences.

Pvt. Baker's memoir:

Friday, November 24

I sent a box home that cost 2.70 francs. It weighed eleven pounds. It had a ring of aluminum made at Arras by a soldier, a Hun belt, a Black Watch kilt, kilt belt with badges and buttons on it, two whale's teeth and other trinkets.

This is one of many references by George Baker to the American penchant for collecting souvenirs. The culture was not limited to soldiers. Laura (and later Erle) collected souvenirs also. Growing up, the Davis children played with them. Even today, a German 4.5-inch brass mortar shell casing serves as a desktop repository for the editor's pens, pencils, scissors, letter-opener and magnifying glass. Laura retrieved it from the Vimy Ridge battlefield after the Armistice. Elsewhere there are her British "tin hat" helmet, her gas mask and her collection of British Empire insignia and ribbons given to her by patients, plus a French helmet from Verdun which Erle brought back, gift of a French veteran friend.

Letter #47

Laura to Erle
France
Nov. 25, 1917.

Dear Buddy:—

Tonight the wind is blowing, the clouds are flying and the moon is hidden and I am busy poking up the fire in our cozy little room. I had a half day and I've spent it reading, sleeping, thinking and poking the fire. . . .

Ward S is no more. Yesterday we moved all the patients out and today we turned in the equipment and the tent was taken down to make room for a hut.

I feel as if I'm out of a job and I had such a queer feeling about tearing it all up. I felt as if the sheriff were selling me out. But I'm glad to shift the responsibility to some one else for a while. I've had the ward for five months and a half now.

It seems so funny to get your letters and read your comment on my mood of two months ago. I'm funny to myself when the moods change so I must be more so to you. . . .

I expect to go on night duty Wednesday night. That will mean I'll sleep over both Thanksgiving and Xmas day. . . .

Tuesday 2 AM

This morning I came over to Hut I ("eye") and went on night duty tonight. I feel very lucky that the sister who was on here became ill because I was posted for two wards. It is blowing and pouring rain tonight so you can imagine the difference between two large wards of 50 beds each and a hut with 30 patients even tho they are sicker. Then, too, it was easy to change to night duty because I've been sort of detached for two days.

This ward is devoted to fractured femurs and the treatment is Balkan frames. Perhaps you know what a Balkan frame is but probably no. It's a bar up over the bed [***drawing of frame within text***] so and a weight is attached to the leg and put on a pulley on the frame. The treatment makes nice straight legs and also makes them easy to dress but you can imagine that the patient is not very comfortable all of the time. They stay in this position for several weeks.

The orderly has come in to talk so if this seems disjointed it's not my fault. I may have to finish up tomorrow after breakfast but I'll probably be very sleepy. I do want to get this in the morning's mail.

The orderly is a smallish, red-haired man with an excellent opinion of himself. Whether he deserves this excellency of opinion I'll tell you later. I decide those things by myself usually without any broad hints from the person himself.

If you could see me tonight you'd think my weight is increasing muchly. But it isn't fat. It's clothes. I am trying to keep warm by the addition of layers. I've added two layers today and am none too hot as I am. By the time the cold weather comes I'll probably look like a Teddy bear.

It hasn't frosted here yet but it's so beastly damp that I think we feel it more than 10° above zero at home. Even so, we are much more comfortably situated here than I had ever thot we would be.

The rain has been pouring steadily since eight o'clock. It's conducive to dreaming or to sleeping under different circumstances. . . .

Later

I've decided about the orderly. I've just found the brandy bottle empty and it was half full when I came on duty. If my patients need stimulation I prefer administering it myself. Also the young man stands to get in trouble because I happen to know that the Medical Officer measured the amount in the bottle last evening because he had grown suspicious.

It is five o'clock and we'll soon begin the morning washings and temperatures. I haven't another thing to say that can be said on paper so I'll say good morning and start to work. I've certainly had an easy night of it. It's been hard to stay awake because of not enough work. I haven't had much time on duty like that so far.

Diary of Dr. Neal:

Nov 29—Thanksgiving in France—dinner with "sisters" (nurses) in their mess hall.

Thanksgiving Day is past and we celebrated in great style at our mess. The officers came up and ate with us. We had table decorations, place cards, menu cards, a six course dinner and a bunch of the poorest speeches I ever hope to hear. But it was a good dinner with turkey, dressing, celery and mince pie. . . .

Pvt. Baker's memoir:

November 29

MENU
For
Thanksgiving Day Dinner

Oyster Soup and Crackers
Roast Turkey and Dressing
Mashed Potatoes with Brown Gravy
Creamed English Peas
Apple Sauce
Celery
Crab Salad
Tomato Catsup
Bread and Butter
Pumpkin Pie
Mixed Nuts
Coffee
Cheese

The above good dinner was paid for by the men, themselves.

Letter #48

Laura to Erle
France
Nov. 30th 191
1:30 AM

Dear Young Man: —

Thanksgiving Day is past and we celebrated in great style at our mess. The officers came up and ate with us. We had table decorations, place cards, menu cards, a six course dinner and a bunch of the poorest speeches I ever hope to hear. But it was a good dinner with turkey, dressing, celery and mince pie. . . .

This is my fourth night on. It has been very easy so far but I'm not very fond of night duty. I'm too stupid while doing it and I sleep so poorly. The first three hours go beautifully and after that it's all up. The hours are quite long on duty, too. We come on at eight and go off at seven thirty. If we could just cut out the hours from 2 to 4

I'd get on beautifully. During those hours I either walk in my sleep or else am semi-conscious and that is no nice way for a person to be.

Bye and bye I'll have to draw this to an abrupt close for the simple reason that I'll not be able to see nor to push a pen.

I forgot whether I have already told you about our new uniforms. They are dark gray gingham with stiff roll down collars and deep cuffs. We all look like Quakeress's but the dresses are becoming to almost everyone. . . .

This afternoon I took a nice little walk all by my self. I walked out toward the sea. There is something very lonely about the scenery here. Pine groves (they're too tiny to be called woods) and yellow sand dunes and low bushes or else spindly poplars with bare branches are lonesome even to speak of, aren't they? It was dusk when I returned and built a fire in our little stove.

We have a new table lamp with a "buful" green shade for our room. We think we are about fixed up for the winter. But something else keeps turning up.

I do hope I get to eat next Thanksgiving dinner at home. I haven't been there for eight years and it is our day to entertain all the family and such relatives as may be within reach. I'll take dinner on Oct 19th ***[Erle's birthday]*** and Dec 25th with you if I may but Thanksgiving is different.

I still hope for the war to end next summer and I keep planning for it to do so. Everyone laughs at me when I say I'm expecting to get home next summer or Fall. I have a hunch we'll get there just the same. How is your hunch on that subject? Now please be optimistic.

I'm not counting on beating up the Germans as they should be but I do hope for something to happen to their war party and rulers. . . .

At that touching point I was interrupted by one of the men on guard duty coming in to see me. He was orderly in my ward for 5 1/2 months and I'm very fond of him. He's one of the nicest boys I've ever known. He's a law student and lived at 541 Barry Ave. last winter. His fraternity had the house. But all this isn't to the point. He is probably the hardest working, most conscientious, most efficient man in the unit—I'd leave out the probably if I knew all the men personally I'm sure—and has been recommended many times for promotion. Because he doesn't happen to be a relative of our Colonel or Major nor because his father wasn't a fraternity brother nor classmate to either of them he is left in the hardest places while inefficient, lazy members of the unit are advanced over his head. Such troubles I suppose will exist in all privately organized units. It's a shame, really. He tried three times to get into Army, Navy, Aviation, anything and couldn't make it because of being under weight. Here he isn't even allowed the privilege of trying to transfer to other services. He'd make a peach of an officer. He can get anybody to do most anything for him and only be happy to do it. And he's simply wild to get up the line.

Perhaps this is all you care to hear of Art Hall but I'm telling you to illustrate conditions in such a "wonderfully organized unit". You should have heard the bouquets they pinned on each others bosoms at the dinner table last evening.

Ooh it's six o'clock & time to take temperatures. . . .

I don't understand at all why our unit hasn't been called into active service. The latest word we have received is that we shall be called about the first of January. I've heard so many rumors, however, and have given serious attention to so many of them, without seeing any materialization whatever that I'm quite cynical now. Until we get real orders to pack up our duds and go to the train I shall continue to take life calmly.

Letter #49

Laura to Erle
This was the last letter received (Dec. 24) by Erle before Christmas and, indeed, before the first of the New Year, 1918.
France
Dec. 3, 1917

Dear Buddy

It's four in the morning. I've just been up to the bunk [office] with my report and then stopped at the second hut from here to see if Conny was busy. She wasn't. She had been making brandy sauce to pour over some fruit cake that her orderly had just gotten in a package from home. And I ate some and talked a bit and ran along back to Hut 1. Can't you imagine us doing such commonplace things away off in France behind the lines while a big convoy of wounded is rolling up to reception tent in the ambulances. We really live along in much the same way that we do at home in institutions. Of course we have to wear more clothing to keep us warm and we must wear nothing but uniforms and our recreations are scarce and limited.

The moon is brilliant. It is high up in the sky with just a few shivery looking clouds floating among the stars (properly: below the stars). And Orion and the Pleiades and a great brilliant star that I don't know are so very distinct. . . .

The Christmas packages have begun to come in—mostly woolen stockings, warm nighties, fruit cake, etc. They make me feel guilty to think that I didn't send some little gifts home but it didn't seem wise to do so.

I do hope your unit will be called. Even if the work itself is very unsatisfactory it will mean a lot to you and there will certainly be the opportunity of writing and of course that is what you want next to your desire to serve our country. After all as long as one serves and does one's best what matters if the organization is not perfect? I suppose it's because I hate inefficiency and graft and petty favoritism that I let such things bother me. . . .

I received the Nov. 10 S.E.P. today. I brot it over to read tonight but haven't had a chance at it yet. There are too many uncomfortable people to look after.

The mornings here if not too cloudy are very pretty. The sun rises behind the hills and tints the clouds so beautifully before it peeps over at us itself. In some ways I like night duty. I love to get out to walk in the morning and I like to lie abed and talk before rising. When working days I can indulge neither of these tastes.

Wed. 1 AM.

I seem to be a failure as a letter writer. Something happens each time and I have to stop. Or else the orderly comes in and begins to talk and I just can't visit satisfactorily with you when anyone else is present.

Today I had a delightful morning walk. The mist was over everything making the trees look feathery and the houses indistinct. The mist was all over the tops of the hills and the yellowish light of the rising sun behind the tops was very satisfying.

"I will lift mine eyes unto the hills" is getting to be my favorite Biblical thot. . . .

Diary of Dr. Neal:

Dec 6—Halifax, Nova Scotia blown to pieces by explosion of munition ship in the Halifax Harbor. 1400 to 1500 killed and thousands wounded and homeless. Ship rammed by a Belgian relief ship.

Erle to his mother

CHICAGO

Friday Evening

December 7, 1917

Dear Mamma,

Your letter certainly was loaded with good news. It sounds like the best piece of good luck we have enjoyed in a very long time. But you haven't a monopoly on it at home.

Last Wednesday the central division Red Cross headquarters called up and asked if I could be borrowed from the Western Newspaper Union for the month of December to help put across the Christmas membership campaign. After I talked the matter over personally with the managers and they had offered me $50 a week to help them, I consulted Mr. Patterson and received his consent.

He asked what I was to be paid, and when I told him, said: "Good; we will contribute something, too. When your month is up over there give me a bill for $10 a week expense account and I will give you a check for $40."

Sixty a week isn't so bad, eh? Be as tickled as you please about the business, but don't tell anyone about the money end of it.

At the end of the fifth day of work at the Red Cross I can say it is a busy job. I didn't finish my stint this evening until 7 o'clock. But hours-of-labor has no consideration with me in the circumstances. The W.N.U. was a loafing job, and I am glad to get a change to real action. The work is very interesting and I am able to do it enthusiastically.

I don't understand at all why our unit hasn't been called into active service. The latest word we have received is that we shall be called about the first of January. I've heard so many rumors, however, and have given serious attention to so many of them, without seeing any materialization whatever that I'm quite cynical now. Until we get real orders to pack up our duds and go to the train I shall continue to take life calmly.

The night after Thanksgiving I attended a very pretty dance given for members of the unit at one of the hotels. Every one enjoyed himself. It was a sort of family affair, for by this time practically all members of the organization, including the women, are pretty well acquainted.

This week I've had three letters from Laura, written the first two weeks of November. She says it is constantly raining and very cold—the penetrating cold of middle February in Chicago. It is the kind that goes clear to one's marrow. She is to have a vacation the last of next February, and if possible will go to Paris and then on to Southern France, and the Mediterranean sea coast. Before she returns to America, Laura will have seen a lot of places and things, I claim.

I might as well give warning now that I'm not going to make any Christmas presents whatever this year. I want to hoard my funds and send as much as possible home when the call comes.

Three weeks ago an executive from the National Red Cross headquarters in Washington came to Chicago and told executives at central division headquarters that an important reorganization of the publicity department at the national capital was about to be made. They are going to send several professional news writers to Europe. Did the local people wish to recommend a Chicago man? They did. His name, they said, was Davis—John Erle Davis of the Western Newspaper Union. I learned of the session several days afterward. So far nothing has come of the recommendation, but something may come of it. Matters move slowly sometimes down at Washington.

I have had but one letter from Aunt Sadie since we were all at Indianapolis. I haven't answered it—though I don't know why I haven't unless it be that my talent for writing letters has dwindled. Some time ago I wrote and thanked Aunt Nannie for the helmet she sent me, but haven't had a reply. . . .

Today I'm all discouraged again. I don't feel quite so sure about the war being over so soon and yet way down deep I'm sure that it can't last forever. And when I'm away from here I'm going to try to forget that there ever existed such a thing as an army or a military hospital.

Letter #50

Laura to Erle
France
Dec. 13 1917

Dear Buddy: —

It's just about a week since I've mailed you a note of any description. I've carried my little basket with letters and stationary, etc. each night to the ward and expected to be able to turn out a few pages. But the work interfered. A man with a basal skull fracture has kept both the orderly and myself hopping and then besides I had an extra ward to look after so—you see—excuses and no letter. . . .

The box of coffee arrived all O.K. in good shape last Monday. Many thanks for it. I shan't scold you for sending it because I do enjoy it so much and in all the packages I've had no one else has thot to send coffee. And I'm sending the return you asked. Do you feel them?

Today I'm all discouraged again. I don't feel quite so sure about the war being over so soon and yet way down deep I'm sure that it can't last forever. And when I'm away from here I'm going to try to forget that there ever existed such a thing as an army or a military hospital.

Miss Urch is due back within two weeks and I'm sure there will be trouble of some sort when she comes. How I do detest little souls in people with some authority. They're bad enough in people of little influence. All of which isn't very enlightening to you—nor to me. I just feel trouble brewing and I'm not a mystic at all. . . .

If that skull fracture man would only keep still a while at night I'd have a chance to read a bit but it's absolutely impossible to do much of anything except work when he is awake. Drugs seem to have no effect at all.

Frances is in the throes of another love affair—the third or fourth or fifth inside of six months and each one worse than the last. She's just bid him good bye. He's going back to England to take a staff position of some sort as he is permanently disabled for service at the front. She has promised not to go out with any one else and we are betting on whether it will last one or two weeks or maybe two and a half. . . .

We've been getting our issue of coal regularly now so we are warm and comfortable and such heaps of Christmas boxes have come that it makes me ashamed. You know we really don't need things here. There's some things we want but they aren't necessities at all.

We are so much more comfortable and better fed than I'd thot we'd be.

We had coffee this morning in our room from the new box you sent. It is certainly good coffee. . . .

Yes I know the Krosts rather well. Capt Krost was the Medical Officer in my ward for the five and a half months before it was torn down. His wife is in charge in the operating room. . . .

Letter #51

Laura to Erle
Dec. 17, 1917

Dear Buddy:-

. . . I didn't get finished yesterday because of the need of my presence in the ward and also because the stove smoked and made my head ache somewhat and then when I had given up in disgust trying to get it to stop smoking I was all black and sooty and my inspiration had oozed away. . . .

. . . at our Thanksgiving dinner one of the majors presented Miss Urch's secretary with a very beautiful diamond. It was quite a surprise to everyone, I think. They had been very friendly but both them are very popular and no one had taken the matter seriously.

You asked about our Halloween festivities. We just had a mask dance at our mess. Yes it was fun. But I had been so busy that day and worked so late that I didn't get to the party till everyone had unmasked. Dancing seems to be our principal recreation. We have a dance of some sort about once a month and occasionally oftener. If we can't get other music we use the Victrola. But I can't claim to have taken much of an interest in any of them. I like to go, dance one or two or three and then to my own room.

Tonight there's at least three inches of slush and the wind is blowing high and cold. It is our first snow and everything looks so unfamiliar with such a covering. And I've been complaining of the cold for almost three months! You'll begin to think I grouse a lot. And I do. But if I didn't what would I talk about? . . .

The hospital is fairly well emptied out just now and probably will be till after Xmas. Unless of course Fritz makes an offensive. That might happen.

My orderly says his mother has it all figured out from the Bible that there will be peace in February. I hope her calculations are right—That is if it is the kind of a peace we want.

We have heard rumors that in May we will move back of our own lines. It is only a rumor but it furnishes us a little excitement any way.

If this letter is all sort of smeared up with coal dust you may know that I'm poking the fire between paragraphs. The coal is poor quality and needs lots of firing but I'm glad to have it anyway regardless of quality. . . .

It's time to wash patients and take temperatures so I'll have to stop now and get to work. I've had many interruptions so many people have stopped in tonight to ask for coffee—boys on guard, a man from the office, the sergeant, the night nurse next door, a friend of mine from the other side of camp and so it goes. Do you think they sniffed real coffee in the air? Oof! . . .

Tonight there's at least three inches of slush and the wind is blowing high and cold. It is our first snow and everything looks so unfamiliar with such a covering. And I've been complaining of the cold for almost three months! You'll begin to think I grouse a lot. And I do. But if I didn't what would I talk about? . . .

Letter #52

Laura to Erle
Dec. 19 1917

Dear Buddy: -

It's cold and crisp and sparkling tonight. There are snow, stars and moonlight. A big convoy is coming in but it will probably be mostly medical ***[sick, not wounded]***. However, convoys effect me not at all in the ward as my patients are stationary and every bed is occupied. Won't these men be delighted to get into nice warm beds tonight? . . .

I'm getting so I don't mind night duty a bit. I'm sleeping so well the last few days trying to make up for the time I've lost before. I only wakened once today and that was when the jar of the antiaircraft guns wakened me at noon. Now please don't imagine that we are being bombed frequently because we aren't. Only once have they dropped bombs in this neighborhood but when the enemy aeroplanes go over the gunners naturally try to get them or frighten them away. I suppose they are mostly scouting or trying to get locations of different things but really don't know. Anyway they come frequently. . . .

We are beginning to get our Xmas decorations ready and at last I'm beginning to get some of the Xmas spirit. . . .

Such a time as I had last night and tonight also with cold feet—not my own but other peoples. I unbandaged and rubbed and applied hot water, hot oil and hot water bags all night long. . . .

Diary of Dr. Neal:

Dec 25—A work day. Christmas for the laboratory in the war zone. Dinner with our nurses—typical American meal. Storm—mixed rain-hail-sleet-snow, with lightening and heavy thunder the late afternoon and evening.

Letter #53

Laura to Erle
France
Dec. 29, 1917

Dear Buddy: —

Please forgive the long time between letters this time. It isn't that I've been busy. I just couldn't write a letter fit to send. I've torn up several beginnings during the week and have finally decided to just go on and talk a while and let whatever I say go in the envelope. I've been so disgusted with the way things have been running lately that I can't think about anything else. I've had all of the army that I want.

The patients had a fine Christmas. The wards were muchly decorated. There was a special dinner, a bottle of stout for each man, music, entertainments, gifts, trees and even a Santa Claus. In Hut I we made net stockings, filled them with nuts, candy, fruit and some little gifts and I hung them on their beds. When they wakened they were like so many kids so anxious were they to see what they had.

Did you know that I'm off night duty? I am. I came off yesterday morning and very much regret that I'll have to start to work again tomorrow. I've slept and cleaned house and made coffee and read a little and have made several attempts at letter writing and now my two days are gone and nothing accomplished at all. . . .

This afternoon Bess and I tried to go to ______ the largest place around here. We got our passes and waited two hours for the train. Then the R.T.O. [***Royal/ Rail Transport Officer?***] informed us that the train would be at least two hours later so we came back to our hut to our nice warm fires. The wind is blowing so hard and is so cold that I'm really glad that we didn't go on. I was intending to purchase a pair of shoes, a shampoo and a large square meal but I'll have to wait another week for all three.

Our room is so cozy tonight. I've had a fire all day and the air is warm and comfortable. I've opened the little doors in front and toasted bread two times today. I love the bright coals.

I'll not try too tell you about our trouble here. I get too mad every time I think of it. Perhaps in another two days I can write better.

Winter 1918, January-March: No End in Sight

THE NURSES REBEL HELPLESSLY

Letter #54

Laura to Erle
France
Jan. 1, 1918

Dearest Buddy: —

Happy New Year! Are you celebrating this holiday very festively or do you work all day? I am doing neither. I am hugging our little stove keeping warm and starting in on a nice long talk to you. Do you realize that I haven't written you a really truly letter for something like two weeks? Of course I didn't mean to neglect you but I just couldn't write for one reason or another. Last week I was too much wrought up over happenings here and everything I started I tore up without finishing. And the one I finally sent wasn't worth sending but I thot I'd better send that than nothing. But I hope it reaches you with some other one so you won't think I'm losing interest. Impossible!!! . . .

I've been off duty a few days with an infection in my eyes but expect to go on soon again. My eyeballs are as red as blood—in fact I suppose there was a slight hemorrhage in the conjunctive and I'm a sight to behold. I'm getting a little discouraged tho'. Three infections in a month and a half are too many. I never used to have anything of the sort. I suppose I probably stick inside too closely now that the weather has become disagreeable. I'll have to change my ways.

I came of[f] night duty last Thursday. I didn't do much on my two days. I had meant to go to ______ the largest town near. Bess and I got our passes, waited two hours for the train and then were told it would be at least two hours longer before it came. So we came on home.

By the end of December 1917, as the Regular Army cadre sorted out leadership qualities of Base Hospital 12's 225 volunteer enlisted personnel—the mostly elite former university students—8 had been promoted to sergeant first class, 17 to sergeant, 9 to corporal, and 96 to private first class. Among the doctors, Gerard Krost and William Harwood had been promoted from first lieutenant to captain. But among the nursing corps, there was trouble.

Stretcher bearers—in addition to scrubbing floors and cleaning bed pans, this is how the privates, many of them from Chicago's best families, were used.

German POWs, also used for stretcher bearing and digging trenches, etc.

It's less than two months till we go on our leave. Did you say that you'd meet me in Paris? Or was it Nice? The girls who have been to Paris say that it is unsatisfactory because so many of the public buildings are closed because of the war.

I still think I'd enjoy a few of the sights there.

Miss Cleveland has been off duty for three days also with a cold and a pain in her neck and shoulder. It has been so very cold that we feel as if we are "swinging the lead" and letting others do the work.

Fortunately we are not very busy in the hospital just now so it isn't so bad. It is very cold in the wards but we manage to get along. It isn't really as bad as I had expected.

Last evening the unit had a big time. The men invited the officers and sisters to a New Year Dance. Everybody had such a fine time. It was the first time since we reached France that we've had anything where the privates were permitted to come. There were about six men to every girl so you can imagine that the ladies at least had a good time.

Miss Cleveland and I of course didn't attend. I was sorry to miss it but didn't care to make an exhibition of my crimson orbs.

Many of the men in the wards have been reading [R]evelations and they insist that according to a certain chapter the war will end in February. Would that I had their faith. I'm never able to read figures into such things and have never been able to believe much in the line of prophecies anyway. . . .

Miss Cleveland read a part of "Mr. Britling sees it through" by H. G. Wells yesterday. I found it fairly interesting but rather tiresome read aloud. I had the feeling of wanting to shove it along. Have you read it? I shall finish it some day by my ownself.

We've been getting plenty of coal recently and these little stoves in our rooms are pretty good heaters. We have a new man to lay our fires and dish out the coal. He is from Sioux City, Iowa and enlisted in the Canadian Army. He came down to the hospital slightly wounded and has been assigned to light duty. He is worth about six of the little Irishman Pat that we had last month. You know I really like Americans heaps better than Englishmen or Irishmen or Scotchmen or Australians or New Zealanders. The Canadians are much like our own people but of course they don't think so. They think they are superior but to me they have the same appearances as Americans.

I can't speak about Frenchmen because I haven't met more than two and I didn't know them well enough to judge.

Red Cross' most recent donation to us has been bright red comforts. I rather like them, they are so warm colored and look sort of homelike on our beds. You know I come from a small town in Indiana and my mother used to always have a red comfort or so.

And as far as our furniture is concerned it is so variable in structure and coloring that I fail to see how a red comfort [can] destroy the symmetry.

But we are comfortable and cozy here and that is the principal thing to be considered in war time isn't it? I wonder will things ever get back to normal? I don't see how Germany can hope to gain much by holding out and yet of course I don't know. This war business is entirely too big a proposition for a head shaped like mine. I can only hope that some unthought of occurrences will bring it all to an end.

I'm going to stop now and get ready to go over to the turkey dinner. Yes, there are turkeys in France. And we indulge ourselves on Thanksgiving, Xmas and New Year's. . . .

I'll not wait so long again if I can possibly help it.

Pvt. Baker's memoir:

January 1, 1918

With the help of the nurses and some patients, we men decorated our Mess Hall for our big blowout last night.

We have in our outfit, without a doubt, the best jazz orchestra in all France. It is led by Jacobs on the violin, with Carsten on the piano. They are sure "there". All of the members are artists, having been students of music in school. They get paid for their playing all over the British area around here.

New Year's Eve was a moderate, clear night and many officers and men had dated up nurses for the evening. I was a free lance, as were most of us. The nurses had their "latest creations" on, which of course, was the talk of the evening. These "new creations" were mostly regular Army issue clothing, but anyway they were real American girls, and, for the most part, very cordial. We had all the dances we wanted but only a couple of the girls were what I'd call good, smooth dancers. These two, Miss Bess Van Ark and Miss Duggan, were sure popular with us regulars. Most of my dances were with Bess who is a particular friend of mine. It was nearing midnight, or around eleven o'clock, before we boys began to loosen up and make everybody at home. I broke all Army etiquette and all general orders from General Headquarters, Army Post Orders #176, American Expeditionary Forces, the U.S. Army in France, by starting to tag a soldier lightly during a dance, to get his partner. We called it the "Roosevelt Progressive Dance". We were in our second dance when I pulled the "Prize Buck" stunt. I had lost my partner and so I goes up gently and taps our C.O. [Commanding Officer], Colonel Collins, on the shoulder and makes him come across with Miss Duggan, our best looking nurse, Ha! Ha! Well, I knew he couldn't reduce a "Buck" very much if he didn't like it and I knew the occasion would be in my favor. But, nevertheless, he seemed a "bit surprised" when he saw me take his good looking nurse away from him. This little incident changed the whole spirit of the evening. From then on everybody was taking the other fellow's partner away. I had broken the ice and had made the old C.O. come to earth and he never said a word. I think he rather liked it because everyone on the floor had seen me and gave a hearty laugh and the C.O. smiled. If a British Buck would ever tap a C.O. on the shoulder at a dance—well, the King would have to step in and save him. I felt that all Americans were equal and on our celebration night, in our own mess hall, we were all equal, officers and men.

Everyone had a good time and today our legs are just aching from dancing.

Heard the noise of an air raid on Boulogne last night also. It was pitch dark which made the star shells and searchlights show up brilliantly all over the sky. As long as Fritzie stays that far away, we don't mind the bombs.

Diary of Dr. Neal:

Jan 4.—Capt. Harwood died of broncho-pneumonia at General Hospital 20, British Officer's Hospital. Buried in Etaples Military Cemetery Jan. 6.

Jan 5—Capt. Dale under diagnosis of active pulmonary tuberculosis—left for U.S.A.

Pvt. Baker's memoir:

Sunday, January 6

We read of the 14 peace points announced by President Wilson.

Went to a funeral for Captain Harwood, our X-ray specialist who died of pneumonia. He was buried with military honors at Etaples. About 150 officers, nurses and men attended. Captain Harwood was an outstanding expert with the x-ray.

Very cold.

Church at 8:30 p.m.

Lieutenant Copp worked on my left eye again. He burned the fine capillaries in the eyeball with a hot needle to clear them up.

Letter #55

Laura to Erle
France
Jan. 6, 1918

Dear Buddy: —

. . . I have very little of interest to tell. Everything moves along in much the same way and nothing has happened since our Xmas trouble.

I don't think I ever told you what it was. I meant to tell you but got so I hated to mention it. I believe I told you before that I almost dreaded to have Miss Urch come back because I felt trouble in the air. The hunch was right. When she returned she was forced to resign as chief nurse and another chief nurse forced to take the appointment. We very nearly had a mutiny. Every nurse in the unit even to the reinforcements who had never seen her backed Miss Urch. We sent committees to the Colonel and sent in a petition and did everything we could only to find that we were up against a wall. "We are in the army now." It is, thru bungling some place, probably in Washington, absolutely in the Colonel's power to do as he pleases in such matters. With a man of Besley's stripe (cheap politician with hypertrophied ego) to stir up trouble this calamity was brot about. Altho if the colonel were a man he would never have done as he did. If we had the commissions we should have had before leaving the U.S.A. we would not be at the mercy of such men. They would have to give a reason not only to us but to headquarters for such performances.

Fortunately for us they, the colonel and the medical director, have chosen an excellent woman for chief nurse. She is an intelligent person, needless to say fairly young, good looking and rather given to social affairs but with a good hard business head. They thot they had an easy mark and a sort of clinging vine but they made a big mistake. She has plenty of sense and "spunk" and will stand for the right things. Of course she isn't Miss Urch and no one else could fill Miss Urch's place in the unit.

The colonel and the major haven't a friend left among the medical officers, sisters nor privates and I'm sure they are aware of it now. They both look like whipped dogs.

Miss Urch is so humiliated. She feels that she is ruined professionally altho I don't see it so myself.

I probably haven't made this intelligible to you but at least it is the main story of the affair. We hope it isn't ended yet.

Did I tell you that so many of the patients have selected a passage in Revelations which they think is prophetic of the war? They read into it or rather interpret from it that the war will end in February.

The report of the refusal by Russia of the peace terms has cheered them up, too. No wonder they are tired of it! They've had a long siege of it.

I had a two pound box of hickory and walnut meats from Silas and the kiddies [***Laura's nephews and nieces***] this week. Wasn't that a nice gift? I do love both nuts and I could imagine all the kids sitting around the table helping pick them out. I'm sure little Alan [***nephew born 1912***] ate more than he picked if he has an appetite like he had when I was home.

The kids are awfully nice if they are naughty. I sometimes think I've missed quite a lot by being away from them so much while they were growing up. But then we can't be in two places at the same time.

How many secretaries, etc., have you attached to your unit? We have two and two other civilians, an interpreter and a dietician and the whole bunch have been superfluous and more or less of a nuisance because not amenable to discipline of any sort.

The interpreter is a Vassar graduate and was a Dean or assistant Dean of Women at Northwestern University. She is a dame of forty odd summers altho possibly not more than thirty nine. She has simply gone man-mad. She simpers and smiles like a girl of sixteen and her face doesn't call for behavior like that. It is long with pointed chin and nose, thin lips, colorless eyes and not a pleasant expression. One of the doctors has remarked that we'll have to hand it to her for getting away with what she does with her face. She toils not neither does she spin. She entertains.

Such is an example of civilians attached to the army. Miss Urch's secretary is quite a charming girl and is well liked and is the only one that hasn't been a trouble maker here. She is the one I told about being engaged to the Major—surgical advisor—and about the Major presenting her with a solitaire at the Thanksgiving dinner.

This letter seems to be mostly gossip but you are fond of gossip, aren't you? If you aren't just drop it in the waste basket and wait for another to come that will please you better.

So you have a hunch that the war will end in the spring? How much is a hunch worth anyway? I've had a few myself and some have come right and some haven't. It's a queer world anyway. The more I see of it the less I seem to know or to understand.

Write when you have time but don't neglect writing to the home folks. You know they are the only home folks you'll ever have but you might find someone else to fill my place in your heart altho I hope you won't.

As you probably notice I'm in a more or less pessimistic mood this afternoon. I think I've been staying in too closely. I think I'll begin to go to all the concerts, dances, dinners, etc. and become a gay butterfly. How I'd love to hear a good opera!

Of course I didn't go anyplace all the while I was on night duty and I've been having trouble with my eyes for a week and have managed to miss every social event possible in our existence here. Don't you think it is about time for me to begin to be giddy? I had never been able to realize before how very much we do need social relaxation when under a strain. I've known for a long time what it meant to the men.

The papers come very irregularly now but I do enjoy them even tho the news is stale. It isn't always stale to me at that. The Post has also been a blessing and Bertha is now sending me Collier's.

The bunch of pictures, I think all from the Red Cross Magazine, were very pretty, indeed. I think I'll have to use some of them to decorate the walls of our "cozy little room". It is cozy even tho you laugh at me for saying so and imagine that it is camoflauge. I've seen many a room in Chicago not a half so cozy nor so home like and since we are issued coal for our little stoves we are very comfortable. . . .

Today one of the girls came in and we talked of marriage—advantages and disadvantages—as we have observed it among our various friends and relatives. We came to the decision that if we couldn't be happier than the majority of married people that we'd much rather stay single and do as we please. But we also agreed that marital happiness all was dependent on the quality of affection.

Now tonight after a walk to the village and a drizzling rain on the roof (it's warmed up a great deal today) I'm yet wondering how in the world one is to gage the depth and quality of one's affections.

I suppose I am thinking more about the subject than there is any necessity but the remark you made in your latest letter concerning your hunch that the war would be over soon and that we'd meet in France, etc. has sort of stuck in my mind. I hope your hunch concerning the war is true. I hope we meet in France but as for the etc. you know that I'm in the army and you may be. If we're both in the army we'll probably never meet here and certainly we'll not marry here. It's too bad to prick the bubble of your idea of the army but, my dear, if your unit is called you'll understand.

I'm so anxious to hear the result of your Red Cross Campaign. The Red Cross has furnished us with quite an outfit but they've sent us some perfectly useless articles. Their most recent donation was a bunch of rain coats. I found one to fit because I happen to be small. Most of the coats were far to[o] small for me so you can imagine how they didn't fit the other girls.

Did I tell you that rumors are rife that we'll move back of the American lines sometime this spring? We still hear rumors but nothing authentic.

I wonder what Doc Flexner thinks he'll use ten wireless operators for? I think they must keep you fed up on hot air most of the times.

We had our first death in the unit since we landed here. Captain Harwood, our Xray man, died of pneumonia and was buried this A.M. It is too bad. He was a nice old chap and worked very hard and everyone liked him. . . .

Excerpt from letter of Daisy Urch to Chicago American Red Cross Nursing Service

January 3, 1918

18th General Hospital, France

…I returned to the Hospital the Saturday before Christmas and walked into trouble. I want you all to know the circumstances so shall write to you in full and trust you will pass it on to Miss Wheeler and Miss Noyes and I hope you will not allow any more units to leave the U.S.A. without the nurses having a definite rank and understanding of their rights.

Let me begin at the very first. Perhaps you know, (as I learned in New York) that Major Besley was to have their uniforms made in Chicago and he kept the telegram from me, so all the money was wasted. Those uniforms paid for by the Chicago Red Cross were discar[d]ed and we were given the proper ones in New York. That was the beginning of my troubles with Major Besley, but not the end by any means. He has made it a point to make it very difficult for me right along. For instance, he would come into the office and demand the nurses stay on duty longer than there was any need, then he would say they were writing things home that they should not. On one occasion he declared that bad sores were developing in the wards. I went all through the hospital and no doctor or nurse could find one. It was one thing after another. He did not like this woman or that woman. When we had guests he would contradict every thing that I said. He began meddling in the nurses mess. I tried to get the commanding officer of the unit to help me, but he said it was Major Besley's unit and he must have his way about everything. It is my private opinion that Besley intended to make trouble for the nurses from the very first. He dropped all his smoothness as soon as we took the oath of allegience.

Well things went from bad to worse. He told the orderlies that they did not have to obey the nurses—so they say. When I returned from my illness he called me into his office and demanded that I resign as chief nurse. I objected, but he stood with his hand on the door and would not let me leave until I did so. What can one do in such a case? Our C.O. says that he is supreme and we can do nothing except through him and he refused to do anything. They have appointed Miss Spencer temporary chief nurse, and I am on duty in the wards. I suppose nothing can be done in this particular case, but surely you can prevent other units from coming over without the chief nurse having her position assumed. If these men can dismiss her at will and without giving a reason, what kind of standards can we uphold? I asked them why and they said they did not have to give a reason and they refused to do so.

Rumor says that Mrs. Besley has been down in Washington working against the nurses getting rank and he has been boasting that she has been working against it. I am so glad that you sent such a good lot of women otherwise do not know where we would have been. We certainly need strong and good women and we need some legislation to protect us. Can you help us?

Very sincerely yours,

Daisy Urch

The Chicago Red Cross office forwarded her letter to the National Red Cross in Washington.

On February 11, the director of Field Nursing for the American Red Cross wrote Dora E. Thompson, superintendent of the Army Nurse Corps:

I am sending an extract of a letter from Miss Daisy Urch, who was assigned to service as Chief Nurse of Base Hospital #12, written to a friend of hers in Chicago. It is an extraordinary situation for Major Besley selected this Chief Nurse himself. It was believed by those who knew both Miss Urch and Major Besley that she had not had adequate administrative experience to justify the selection, and several people who were interested in organizing of the hospital as well as this office called Major Besley's attention to the fact that a nurse of experience should be chosen. As Miss Urch was his selection it seems incredible that he should treat her with so little consideration.

This plea, however, fell before the affidavits of the top two officers of Base Hospital 12:

Statement of Colonel C. C. Collins, Medical Corps, U. S. Army

Reserve Nurse Daisy D. Urch was never appointed Chief Nurse of this hospital. She was appointed Temporary Chief Nurse under authority contained in Par. 61, Manual of The Medical Department, 1916 on the recommendation of Major Frederic A. Besley, the Director of the Unit when I took command. She was relieved from duty on December 24, 1917 under authority contained in Par. 61(a) which reads:

"Temporary Chief Nurses may be relieved from duty as such and assigned to duty as nurses at the discretion of their immediate Commanding Officers."

for the following reasons: She did not adapt herself to the conditions of service in a military hospital or show the necessary executive ability. She was found lacking in good judgment and tact; she was disloyal and showed a disposition not to cooperate with the Commanding Officer. She was found to be temperamentally unsuited for the position she occupied. Every effort was made extending over a period of nearly seven (7) months to secure her cooperation and loyal support but without success. At first her attitude was believed to be due to her ill health; she was put on sick report, transferred to a British hospital for nurses and later sent to Southern France for a period of recuperation. During her absence a marked improvement in the attitude of the nurses on duty here was noticed. On Miss Urch's return she showed plainly by her actions and words that she could not serve in her capacity as temporary chief nurse and give me her loyal support and cooperation. I then explained the situation to her and asked her to resign, which she refused to do. I then issued an order under the authority quoted above relieving her from duty as temporary chief nurse and assigning her to duty as nurse. Her work as nurse has been satisfactory but she still maintains an attitude of hostility to me personally.

(Signed) C. C. Collins,

Colonel, Medical Corps, U. S. A.

Statement of Major Frederic A. Besley, M. R. C., U. S. A.

As Director of Base Hospital No. 12, U. S. A., I had selected Miss Daisy D. Urch as Directing Nurse and when the unit was called into active service I recommended to the Commanding Officer that she be made temporary chief nurse. On the trip over, before we landed in England she gave evidence of being devoid of tact and she was disloyal to and critical of the Commanding Officer. When we arrived at this hospital and took up the work she continued to be disloyal to the Commanding Officer and to manifest antagonism to any directing or controlling authority. After she had returned from a prolonged leave I heard her inform the Commanding Officer that it was her desire not to have any direct communication with him but that she proposed to continue acting as temporary chief nurse and consult him only through her appointed assistants. At no time did she display any aptitude for the position and her attitude was always distinctly antagonistic. I advised her removal.

(Signed) Frederic A. Besley,

Major, M. R. C., U. S. A.

Nevertheless, the officers' classification later of Miss Urch as merely "temporary chief nurse" or "directing nurse" are disputed by an official War Department document dated May 14, 1917, clearly identifying Daisy D. Urch, #5981, as chief nurse of Base Hospital 12 on the eve of its departure for Europe. The logical conclusion is that a clash of personalities had arisen, to Miss Urch's downfall in the absence of established procedures for review and the general lack of official standing held by the Army nurses. In addition, none of the volunteer personnel of BH 12 had experience with Army ways and discipline prior to embarking for Europe.

Army nurses were not granted military officer commissions until 1922. In a U.S. Army hospital in the year 2017, a chief nurse like Miss Urch would rank either as an O5 (Lieutenant Colonel) or O6 (Colonel) and a head ward nurse such as Laura as an O4 (Major). My observation is that the Suffragette Movement among newly aware women professionals such as graduate nurses fueled the ire of the Base Hospital 12 nurses as much as their perceived injustice of Miss Urch's removal.

The 19th Amendment establishing voting rights for women was adopted August 26, 1920.

Snow covered Quonset hut.

Winter Battle Respite

Pvt. Baker's memoir:

January 7

Very little fighting in this cold weather.

Letter #56

Laura to Erle
Monday [Jan. 7, 1918]

My dear Buddy:

. . . Today I went to work and was so happy to do it. I had grown so tired and depressed by just waiting to be allowed to go to work. My eyes are almost well. One is clear and the other is making rapid progress. . . .

The Tommies and Jocks are such funny folk. The wards aren't at all busy and I spent some little time sitting by the fire talking to them or rather listening to them talk. They discuss most any subject so freely. They tell about their love affairs just as naturally as children might do so. And, by the way, did I ever tell you of the very frank love making in England. They walk along the streets with arms around each other and hold hands just as publicly and as conspicuously as possible. We were quite amazed at first at some of the demonstrations but I'm sure they weren't meant for our edification. They were simply unconcerned about outsiders.

They are all so tired of the war and anxious to get back to their wives and babies. You'd enjoy hearing the Jocks talk. They are surely "set in their ways" and all very firm in the belief that Scotland (they say Scotland sounding the o as ou in ought) is the "bonniest country in the world." You'd have to hear them talk to appreciate the dialect. The Englishmen also talk very differently than we Americans do but unless he is a Yorkshire or Derbyshire man I have no trouble to understand. I had no idea there were so many dialects in England. They are very marked.

Tonight I took over the last bunch of picture papers that you sent. They usually enjoy them very much. . . .

Yes, I think you have the correct memory of Miss C. but I'm not becoming so intimate with her. I room with her and am thrown with her a great deal. She doesn't touch me closely, however. She is quite self sufficient and quite well satisfied with herself—both of which are admirable qualities but not conducive to extremely close friendships. But as far as that is concerned I never expect to have a very close friend again. . . .

I don't feel that I've been cheated. Not every one has had one real close friendship stretching over a number of years as Edna's [***Tech, who evidently had died of disease in 1916***] and mine did. . . .

Tuesday

Today we are having quite a blizzard but nothing like they must be having in Chicago. I see by the papers today that you are quite snowed under. . .

Wednesday

I moved again today. I went just across the road to another ward much like the one I was in for the last two days. Again I was not busy so had to put in my time waiting with the Tommies. They are an interesting lot. It has been snowing all day but it isn't so cold as yesterday even tho we broke the ice on the pitcher of water this morning to wash. It seems to have warmed up a bit during the day.

. . . Did I tell you about the big Jock that insisted upon calling me "Bonny Wee Lassy" instead of the dignified "Sister". . . .

Pvt. Baker's memoir:

January 8

Teeth cleaned by Dr. Clark, our dentist.

Guard tonight. I picked a peach of a night to be out walking for guard duty. Very cold and windy. Had coffee in Hut #3 at twelve midnight with nurse and night orderly. I pinched some coal for Hut F. I'm supposed to guard coal. It is so precious. I took a chance. I would have frozen if it were not for those nurses and men on night duty.

The boys up at the line are all dug in for the winter. They come in, a few every day, with trench, or frozen feet. The Scots [wearing kilts] do not even have their knees covered. When they get wounded in the knees it makes them look pretty bad.

Diary of Dr. Neal:

Jan 8—Maj. Besley left for special duty at A.E.F. Hqtrs., France. *[In light of the uproar around the dismissal of Miss Urch, one has to wonder if Besley's new assignment was Col. Collins' way of defusing the situation.]*

Jan 8 to 20—patient [self] in British General Hospital with bronchitis—broncho-pneumonia.

Letter #57

Laura to Erle
Jan. 14, 1918

My dear Mr. Davis: —

It's a lovely war, isn't it? For at least one half of my time on duty today I've sat beside the stove in a group of Tommies and Jocks and talked but mostly I've listened. I'm terribly fond of them and I love to hear them talk.

Today I've been up in the trenches, over the top; I've seen deserters shot, I've seen Johnny coming over with hands up crying "Kamerad!, Karmerad!"; I've been to the Dardanelles watching good soldiers dying like flies with dysentery and their bodies thrown in piles and burned. Then I've watched the battalion doctor looking over the men and giving a man with a sprained ankle "Medicine and duty" and telling another he was "swinging it" and then found his lifeless body on the floor of the bath house within an hour. I've even been across to Bonny Scotland and watched the mothers of lads who will never return flock around the one who has returned asking for information about "last words", the burial, etc. and have watched the brave Scotch lad lie manfully about his comrades graves. I've admired the photos of fat babies, cunning (meaning hugable) youngsters of two or three years and sad looking wives and mothers. The wives are always sad and worn looking. Today almost every story was a tragedy. Yesterday it was comedy.

Do you wonder that [I] like to sit and listen? It's horrible and yet all so intensely interesting. It is always a mystery to me how some men can go thru what they do without a scratch. Some have been in service since the beginning and over the top many times and yet seem to escape everything in the shape of bullets and shrapnel.

Today about half the nurses took their first injection of a paratyphoid serum. It is a French product and is not so pure as the American make. It makes almost everyone ill. I don't get mine till Wednesday.

I am in a light surgical ward and we have very few bed patients. Conny is in charge. It is such a relief to not be responsible for everything but just to go ahead with the beds and dressings and cleaning and let someone else worry about supplies or whether this Tommy paraded before the colonel at 8:30 AM or whether that Jock drew his kit from the pack store at 2 PM or whether the orderlies emptied the dressing buckets etc. all day long.

But I'm having almost too lazy a time. We've almost nothing to do in the afternoons.

A few days ago I had a nice letter from your Aunt Sadie. I did appreciate it so much but it did amuse me too—not the letter but her idea of our work. You know there's nothing heroic about our work. Our hardships are mostly imaginary and I think practically all of us enjoy the work very much. The worst part is in being cut off from dear ones at home. There isn't much to do socially here and that is a hardship for some of us. But then we didn't or shouldn't have come with the expectation of shining socially.

Rumors keep flying about that we are to move in the Spring back of our own lines. We await definite orders however. I doubt if we go before May.

I wonder if you'd think me a perfectly dreadful person if I asked you to send me something? No, it isn't woolens. I'm plentifully supplied with clothing. It's something I don't need at all but just want. All day I've been longing for some stick candy—not

Today I've been up in the trenches, over the top; I've seen deserters shot, I've seen Johnny coming over with hands up crying "Kamerad!, Karmerad!"; I've been to the Dardanelles watching good soldiers dying like flies with dysentery and their bodies thrown in piles and burned. Then I've watched the battalion doctor looking over the men and giving a man with a sprained ankle "Medicine and duty" and telling another he was "swinging it" and then found his lifeless body on the floor of the bath house within an hour.

the dainty "opera sticks" but real hard, old fashioned stick—clove and cinnamon. Would it be too much trouble to send me some? Some companies put it up in round tin cans holding perhaps four lbs. I'm rather vague as to weights but I can see the packages distinctly in my mind. I'd ask Bertha to send it but you know that they don't carry it in tins in small town stores. It's all in buckets or else in paste board wrappings and would be all sticky and half spoiled when it came. . . .

Now if you aren't in Chicago—you may be in France already for all that I know—or if it's too much inconvenience don't think of bothering with it. Just pretend that you didn't receive the request and I'll be none the wiser and no worse off. Candy is very hard on the teeth anyway. . . .

I rather (pronounced rothah) think you imagined a good part of what you read into Helen K.'s letter. I didn't see it. But Helen has a heap of intuition. Perhaps she feels something in the air. Yes, I'm sure the girls here have all a pretty clear idea of the state of our minds or should I say hearts?? . . .

[*On a separate sheet of stationery, folded around a clipping*]

I'm so anxious to hear of the outcome of the Xmas campaign for R.C. members. Is this one of your write ups? Bertha enclosed it—evidently cut from the "Sun" or "Plain Dealer" [***North Vernon newspapers***].

This typewriter paper was sent to the I. T. S. [***Illinois Training School for Nurses***] graduates by the Alumnae Association as a Xmas gift. I was surely glad to get my share of it.

The clipping, a display advertisement, reads:
BACK THE MEN WHO FIGHT FOR YOU

For the United States war—real war—has not begun.

Not before late winter or early spring will Americans begin to experience the pangs of grief that have been France's, England's, Canada's, for three fearful years.

Are we to wait until long lists of casualties arrive before we give a vote of support to the world's greatest mercy institution—the American Red Cross?

Countless thousands of troops in cantonments here and in camps "over there" know that their Good Samaritan, the Red Cross, this week is out to enroll 15,000,000 folks back home as Red Cross members.

Picture the encouragement these fighting men will experience when they know that this great goal has been attained. Better still, visualize the picture we can make for them if we furnish a background of not merely 15,000,000 but 20,000,000 or even 25,000,000.

"By your Red Cross Service Flag we will know you."

When you join this great civilian army of the Red Cross, by taking out a dollar membership, our local Red Cross chapter or branch will supply you with a Red Cross Service Flag. Hang it in your window.

Every home excepting those that are pro-German; every home excepting those that are for the "Meund Gott" Kaiser and his cruel, hideous crimes; every house save those that shelter traitors will have a Red Cross Service Flag in one of its front windows before Christmas Eve.

Pvt. Baker's memoir:

January 19

Inspection of whole detachment by 1st Lt. F. A. Harrison who has just been promoted from Top Sergeant. We call him, "Old Humpback". He has about eighteen years service on his record—a fair fellow at that.

Diary of Dr. Neal:

Jan 20—Miss Hinton, a nurse, died from epidemic cerebro-spinal meningitis.

Pvt. Baker's memoir:

January 22

Funeral for Miss Hinman [Hinton], a nurse, who died of spinal meningitis. Buried at Etaples. Chaplain Dancey presided at the grave…

Received convoy of one hundred sixty men last night at 8:00 p.m. and carried wounded men for nearly two hours. This was nothing unusual, just "carrying on."

Letter #58

Laura to Erle
France
Jan. 23, 1918

M'dear Friend: —

It's been so long since I have heard from you that I don't feel justified in writing a more personal beginning to this. Twenty days have passed since I've had a letter in a large envelope with a printed address. Last night a big mail came in. I had letters from most of my relatives, Helen Kelley, Miss Stodola and several Xmas cards. It's nice to have relatives and friends, isn't it? But they aren't everything in the world after all. There's always an empty space that needs filling in and just now it's most horribly vacant because I didn't get just the mail that I wanted.

One time I promised to write a letter and not mention the mails. This isn't that letter. . . .

Yesterday morning we buried one of our nurses—spinal meningitis. The burial gave me the woolies. Everything went off nicely but that terribly bleak bare cemetery with the rows and rows and rows of little brown wooden crosses was almost too much. I'm glad that Edna went as she did instead of as this girl did.

Today I heard that we will probably move within sixty days. We don't know where we are going but we'll be back of the American lines. Perhaps after all we'll meet in France some day. I wonder would we recognize each other!! Queer world!

Frances just called me into her room to tell me that she is engaged to a Canadian officer (Business of being surprised. We've all known it for a month) and to read me his last letter. She couldn't keep a thing to herself if her life depended upon it. I didn't want to hear his letter but she read it anyway. I wonder how long it will last—another month or two? But as I remarked before—Queer world!

There certainly is a large variety of people on this old ball. I couldn't read one of your letters to anybody—I just couldn't. And anyway what are one's own personal letters to an outsider.

I'm in the same ward as before but we are somewhat busier. We've had a good many convoys and have evacuated very few cases so we are full up. Conny is off duty and I have to manage her ward and I don't like it. I don't want ever to have to manage a ward again. It's so much easier not to do it nor to worry if things are going right or not. I suppose I'm getting lazy.

We are having such delightful moon-light nights now. It isn't cold this week and a walk down to the beach and back would be tres bien, would it not?

I had such oodles to say when I sat down here and now I can think of nothing more except that when that letter arrives, if it ever does, I'll be very glad to see it! I lose interest in writing when I don't get an answer.

Honey, you know surely that I'm grousing at the mails and not at you. I know you've written and that the letters will show up eventually. You'll think I'm nothing but a grouch. In reality I feel quite cheerful. . . .

Pvt. Baker's memoir:

January 24

We try to keep warm in the daytime but at night our tents, which have room for three Army cots, are sure like ice houses. (I never thought I'd ever put in a whole winter in a tent with no heat, just a floor of boards. One can get used to anything, I guess). We close the flap on the bell tent and keep the snow out. But, wow!, is it cold to get up at Reveille in the dark, or to carry in wounded men who come in at all hours of the night.

The Scots wear kilts and their legs are bare from below their knees to six inches above the knees. The Gurkha Indians, a fierce looking type of black man, were all sent to the Near East to fight for it was too cold for them up here. It was too cold for the Jamaican, Central American Negroes also. They are a dark-skinned race, highly educated, and gentlemen, but the most homesick of all the troops in France. I felt sorry for them. They did not know what the big show was all about.

We have Welshmen, who are mostly miners, and Moro Indians from New Zealand, another dark-skinned, intelligent race. It is interesting to hear all these boys tell of their home countries.

Sunday, January 27

Half day off. Fine day. Took walk to hills with Ray Fabian. More good letters.

Church, 8:30 p.m. Communion Service, Dr. Dancey preached. A half dozen boys and eight or nine nurses stayed for the whole service.

I was chosen Captain of one of the new Orderly Basketball Teams. We played one game and it was so rough we nearly tore down the Mess Hall so the Colonel cut all further games. Everyone had to fight, trip or bang up against opponents to keep in the game. We were all bruises after game. I never played in any game that was half as rough as that one.

Letter #59

Laura to Erle
France
Jan. 27, 1918

Dear Buddy Boy: —

. . . I had a half day and had taken a long walk all alone. I went down to the next village to our own and inspected the old, old church with the Roman font and the damp musty smell. Then I walked home again and had no adventures at all till I reached the P.O. There I heard that twenty bags of U.S. Mail had just arrived and came on to our hut feeling so expectant. . . .

A clipping pertaining to a Chicago heat wave in late February 1918 was enclosed in the envelope as opened, but obviously came from a later letter, placed here because of the reference to mild weather on the Channel coast in January.

The weather has been like May for the past week. The air is almost balmy and the moon is "tres bien". (I notice you speak French so I'll try to work off my ten word vocabulary on you) The atmosphere is clear or else just misty enough to be beautiful. I so wish you were here to enjoy it with me. I'd much rather be walking with you tonight than writing.

I hear "men" being discussed outside my door. I wonder if you know how little the subject means to me? They're so uninteresting that I refuse to get excited about them. And they may be picked up on any corner. Even the most unbeautiful one of us may have beaus in plenty if she so desires. . . .

I told you, I'm sure, that rumors keep flying about concerning our moving back of the American lines? Gossip now says that we'll go within sixty days. I've heard that we are going to the muddiest place in France. There are so many "muddiest places in France" that the information received is rather indefinite. I hate to think of packing up all our junk and also of again furnishing up a room to make it inhabitable. Your idea of our room makes me "laff" every time I think of it. And it is cozy and quite home-like especially since we've purchased a lamp with a soft green silk shade. Bertha makes fun of our red hangings in combination with me and my coloring. But you see red matches my roommate and that makes it perfectly all right. Red is rather a warm color any way.

You know, honey, I've been wanting to write you every night this week. But at first I was waiting for mail and wouldn't. Then Friday night I went to a little dance and had a very nice time. I was so happy over the letters that I was quite gay for me and danced every one even after the lights went out (air raid some place in the district) and we put up the shade and tripped by the light of the moon. Do you remember our moon light dances at 453 Belmont Ave? But of course you do. Then last evening I went to a so-called concert at the Y.M.C.A. hut and was bored stiff wishing all the time I'd stayed home.

I find it's bad for me to stick home too closely and I'm going to go places whether I specially want to or not. At least they take me away from thoughts of the work. How I'd love to go down to St. Mary's with you and listen to the Choristers.

You made no mistake in telling Miss Flexner that I enjoy her verses [***Hortense Flexner was a pioneer woman journalist and one of the famous Louisville Flexner medical family, later an author of children's stories and professor at Sarah Lawrence College]***. I do like them as well and also Dave Morton's ***[Another Louisville journalist and later Amherst professor known for his sonnets***].

Everything you've sent me by either of them is so artistic and so beautiful. In one or two I haven't quite caught the thot because I'm so earthy, I suppose. But usually I do. They sort of take me away from the plodding world into a dream. It must be wonderful to have such friends. Perhaps some day I'll meet them and they'll wonder what you see in a little mouse. I often wonder my self about the same thing. In fact the whole thing is full of wonder to me. I didn't ever think I'd be so foolish about anyone nor care so much and so long. But I'm very glad to do it, aren't you? Life would be so dull and uncolored without such an experience.

Conny's roommate is camping over here with me while Miss C. is quarantined in Conny's room. She is such a pretty girl. Her complexion, eyes, hair, mouth, rosy cheeks, tall slender figure are not at all hard to look at. If I had been constructed like that I could understand a bit why you care. Isn't it strange that some people draw all the beauty, nice dispositions, charming personalities, etc., and leave only scraps to be divided up among all the rest?

When I used to talk so my mother would say, "Well, Laura, if you are only as good as you look you'll do very well," and she wasn't complimenting my looks either. She was meaning something very different from flattery. . . .

You speak of air raids as if you thought we served them for tea. They are very rare here—in fact they've bombed this immediate vicinity only one time since we've been here. Of course we get the signal even if they're forty miles away and can always "get the wind up" if we so desire.

Two days ago I received another package of coffee, mailed I suppose in early December. Only about one half a lb was spilled out this time and some one had accomodatingly jammed a box of cocoa down in the empty space—probably from another broken package. It is very thotful of you to send it. I can't write you my appreciation but I feel it. Please don't send any more coffee now. . .

I also had a roll of papers from you this week. I haven't had time to read them yet but I shall soon.

I'm glad you enjoyed "Christine." I was sure you would find it as delightful as I did.

I read "Four Days" when it came out in "The Atlantic Monthly." I enjoyed it but it left me sad—I suppose I'm getting like the soldiers in my demand for cheerful literature. There's too much of tragedy of all sorts to be seen and heard here.

There is "Taps". We don't observe the "lights out" ruling but I'm afraid I'll tire you with this patter. . . .

Pvt. Baker's memoir:

January 29

Half day off again. Parish, my ward partner, and I have arranged it this way. One of us gets off at 12:30 p.m. every other day.

Light duty men of the hospital were "marked out" and Smith, one of our patients who had recovered, went from our Hut F. He is a fine fellow.

Warm bright day. The fellows, feeling a little touch of spring, tossed a baseball around a little.

Everything rather quiet at the front now. We cannot get much news from there. Both sides are preparing for spring drives.

January 30

All men, officers and nurses heard Major Lindsley speak on the new war risk insurance. I took out another $5,000 so I now have $10,000 U.S. War Risk Insurance. If I get disabled or killed, Mother and Dad will receive $57.50 a month for twenty years.

January 31

Muster, meaning we line up for our pay, for January, and inspection of the whole detachment.

Colonel Collins just returned from Paris. He is a full fledged Colonel now with silver spread eagle. He shook hands with every man and we all congratulated him. He is one of the most immaculate officers in the United States Army.

Letter #60

Laura to Erle
France
Thursday, Jan 31st

Dearest Buddy: —

. . . Yesterday I had such a good time. I had a whole day off—the first since last August—and I spent it in gadding and eating, etc. Bess also had the day so we secured passes on the train to a neighboring town or perhaps it's a city [Burlogne]. We got off before we reached that place and took a tram out across the hills and looked at an old castle of Henry VIII fame. It was picturesque but I don't believe I'd like to live in a castle.

Then we took the tram back again to the city and wandered about trying to find ourselves and do a little shopping. After a while we got things located, a few land marks memorized and then we saw some of the sights. We climbed the hill to the old city wall and walked along the top of it. I don't know how old the wall is but it certainly looks just as one would expect a city wall to look except that it is much thicker than I had thot. Trees and grass grow on top and it makes a very nice walk.

We went thru the old gate or rather where the gate used to be and wandered around a bit and found the cathedral. The cathedral is surely beautiful. It far surpasses St. Paul's in London. The decorations were very satisfying and so was the architecture. A circular chapel in the rear under the big dome was wonderful. . .

We're still in doubt as to our location when we move or if we move at all. Also we hear that leaves to Paris are stopped. So perhaps I'll have to ask you to meet me in Nice or some place on the Riviera wherever that may be.

Tonight we are having another little dance. The M.O.'s are giving it and the staff of a Canadian hospital nearby is invited. I'll go up and do a few Pig rassels just for fun. . . .

Yesterday I received from the mother of one of my patients the nicest little plaid bound edition of "Lady of the Lake". I am so pleased with it. And the patient is the dearest young Scotchman. I think he is a University student altho I'm not sure. Just now he is serving as a barber. I'm very fond of him and so is Conny. I'm afraid we spoil him. If I had a younger brother I'd like him to be like Jimmy. . . .

Diary of Dr. Neal:

Jan 31—ordered to report to the A.E.F. Central Medical Department Laboratory, Dijon, Cote d'Or, France.

Feb 2—Left Base Hospital No 12 for Dijon.

The girls who have returned from vacations report a very wonderful time in Paris. I'm getting so anxious to go. Wouldn't it be delightful if the gods had so ordered things that you could meet me there? I wonder some times why so few delightful things come to pass. It must be because we don't deserve them or perhaps because we might not be able to stand the test.

Letter #61

Laura to Erle
France
Feb. 3, 1918

My dear Buddy Boy: —

. . . I have been so worried since I learned of Silas' injury ***[A mail clerk on the Baltimore & Ohio, he was injured—and permanently disabled—in a railroad accident between North Vernon and Louisville in January 1918.]*** I missed Bertha's letter giving details and all I know is that he has a fractured arm and leg. Of course that might mean most anything, there are so many different kinds of fractures. I'm waiting eagerly for more news concerning him altho Bertha said he was getting on nicely. . . .

Mon. Evening

I didn't finish last night because of my third inoculation of Para-Typhoid Serum and to-night I was worried for fear Miss C. would stay home from the lecture at the Y.M.C.A. She almost gave it up but I urged her on by telling her it had been a long time since we'd had the opportunity ("So sorry that I feel too rotten to go myself"). . . .

Tonight there's just a tiny patter of rain on the roof and our little coal stove is glowing and the tea kettle is singing. I don't hear a voice in the whole hut. . . .

Tonight I received a "compliment" which was exceedingly funny to me and to the girls here who know me so well. But it may cheer you up a bit. (!!!) In the ward Miss C. mentioned me to one of the Drs. He smiled and said "Oh, Miss Huckleberry. She has always seemed to me to be an ideal little housewife."

I might add to this anecdote that what he knows about my housekeeping or me, in fact, is that gained thru casual evening calls at our flat. He was a sort of family friend to the bunch of us.

The work of the whole hospital is very pokey just now and uninteresting. We have fairly frequent convoys of medical patients but seldom get very sick ones. The only interesting feature is the men's conversation and I'm sure I've relayed some of that to you. Since my day last week when I visited the old castle and cathedral I haven't been out of camp and nothing has happened in camp. We've heard no more news of moving.

The girls who have returned from vacations report a very wonderful time in Paris. I'm getting so anxious to go. Wouldn't it be delightful if the gods had so ordered things that you could meet me there? I wonder some times why so few delightful things come to pass. It must be because we don't deserve them or perhaps because we might not be able to stand the test.

Perhaps some day some more delightful or beautiful things will really happen. But now we hear so much of the war and atrocities and graft and spies, etc. that I almost hate to look at a newspaper . . .

First Leave and Sightseeing in Paris

Letter #62

Laura to Erle

The typical tan envelope, sealed by censor #3840, was date-stamped RECEIVED 3:00 MARCH 18 1918 WESTERN NEWSPAPER UNION CHICAGO. By this time, Erle had received his orders March 4 to report for active duty April 2 at Camp Dodge, Iowa, with U. S. Army Base Hospital 11. Unit 11 was organized May 1, 1917, from Erle's Illinois State Guard unit with professional staffs from St. Mary's of Nazareth, St. Joseph's and Augustana Hospitals of Chicago. Thus they had considerably more experience with Army requirements than Laura's BH 12 had had eleven months earlier.

France

Feb. 7 1918.

[No salutation]

Honey, did you know I'm going to have a leave and it's going to begin next Monday and I'm going to Paris? It's a fact and I'm all excited about it. I haven't been before because it seemed rather far away. But because the leaves have been shortened from two weeks to one week with two days allowed for travel our time comes sooner.

Next Monday night I'll be in Paris. If you could only meet me there! But the Fates have decreed otherwise. If they'll just decree a finish to the war soon I'll forgive them much of their misdoings in the past. There's so many interesting, beautiful and useful things to do it seems rather a shame that the whole world must stop to help put one kaiser and his duped subjects where they belong. Can you just think of millions of things you want to do apre le guerre (spelling wrong I'm sure)? I keep thinking of something new every day. . . .

I've just come from the mess where some very good Jazz Band records are playing on the Victrola and I'd love to dance but I didn't. . . . Then, too, I had on my rubber boots and they aren't the most satisfactory dancing slippers in the world. . . .

Just now we seem no nearer the end and yet I can't help but feel that we are nearer than anyone knows. I can't exactly remember any hunches that I've ever had that came true but just the same I feel that some time this year I'll be coming back across the Atlantic. That hunch stays with me in spite of the fact that every one is preparing to stay indefinitely.

Emma, Bess and Ella have just gone to the Y.M.C.A. to the movies. Frances staid home to write letters and so did I and they all seem to think it's a huge joke. I didn't mention they just "guessed". Miss C. is on duty but will soon be in again. She almost went on night duty tonight except that we go on leave so soon.

I've had two lbs. of coffee from home during the past two weeks and several boxes of chocolates. Did you think I had gone completely mad when I asked you to send me some stick candy? 'Tis no wonder if you did. It was surely a childish thing to do and one of those impulses that should have been considered before putting into action. Wouldn't it be delightful if some time I'd learn to be a "perfect lady" and never do anything rash nor rude nor uncultured and never have to "wish I hadn't". But you wouldn't recognize me then as an old friend and fellow citizen of North Vernon, Indiana, U.S.A. . . .

Did I tell you that Frances, Miss C. and I are going to Paris but that the other three, Polly, Miss Matzen and Bess are going to Nice instead. They say they want a change of climate. But just supposing that we don't get another leave. It would be too bad to be in France and not get to Paris, wouldn't it? I can imagine how peeved the Fritzies must feel. . . .

Frances' case is still on. No other man has loomed upon the horizon and she seems to have all the well known symptoms. Wouldn't it be strange if it really lasted? But she has told me of so many who were "the only ones she ever really loved" that I'm a bit skeptical. However, I'm open minded and willing to give her the benefit of the doubt. We're all built so differently that we just can't be alike. I'm glad we aren't all alike anyway. Think what an awful monotony life would be.

Pvt. Baker's memoir:

Saturday, February 9

Received my first pass to gay "Paree" and have made plans to go for ten days with Oscar White of the Boston Unit, a buddy of fine character.

Letter #63

Laura to Erle
Feb. 9, 1918 [Saturday]
[No salutation]

Yes, Buddy, I'm really going on Monday! When I used to know you did you ever think of me as going to Paris? I'm sure you didn't. Altho all things seem possible when we are in high school I don't think I ever expected to reach this side of the Atlantic. . . .

Mac just arrived this P.M. from her leave and she reports a wonderful time and Paris half full of Americans and no trouble to find one's way about the city. . . .

This evening I've been browsing over the little volume of [James Whitcomb] Riley verse. It is sweet and wholesome and restful and dreamy and maybe that's why I wanted to talk to you tonight. . . .

Sunday Evening

. . . I've just been in seeing Mac and finding out all there is to see in the gay city. Now I'm going to repack and get to bed . . . .

Pvt. Baker's memoir:

Monday, February 11

Left Camiers at 2:00 a.m. on the Calais-Paris train and arrived in Paris at 8:25 a.m., after about six hours on the train.

I got on second class at first, then dropped off and piled on third class because all the seconds were packed with Australian boys going the same place on leave. After a few more miles I dropped off again at a station and ran ahead and got on a first class car where I stayed the rest of the trip. The conductor had collected first class before I came in so I got to Paris for nothing. Not bad you know! In France one cannot walk from one coach to another as we do in America, as the doors are on the sides of the cars rather than at the ends. In order to move from one car to another, one has to jump off one car and run over to the other. This, of course, is done when the train has stopped for some reason or other.

I met Oscar S. White of the Boston Unit who was to see Paris with me. There were three other Harvard boys there, too.

Oscar and I went to the "Y" Hotel, (Hotel Pavilion), 36 Rue de Elychique, where we took rooms on the third floor for 8 francs a day or 4 francs each. It was a fine place and had a large parlor and lounge room and a nice dining room. We met many U.S. boys here. They wondered how we had secured chevrons for six months service already.

We went out sightseeing on the main boulevard to get the lay of the land. We had our first ride on the Metro, or Subway, when we went after tickets for a show. We got tickets which cost 7.5 francs each for good seats at the "Theatre Femina". This is a beautiful theater and we saw a fine up-to-date performance. The whole show was a display of real Parisian life, beautiful girls with gorgeous gowns, etc. I was on the verge of falling asleep as I was so tired from traveling and going strong for about eighteen hours.

We hit the hay about 11:30 p.m. and were sure tired. After eight months on army cots, those beds seemed wonderful to us and we slept like logs. The covers were so deep we could get lost in them. If I hadn't awakened each day and dragged Oscar out we would have spent all day there. We managed to get down to breakfast at nine or so and were just about the last ones served. Our breakfast consisted of two eggs, bread, butter, jam or jelly, coffee, oranges and apples, costing 2.75 to 3.5 francs.

We hit the hay about 11:30 p.m. and were sure tired. After eight months on army cots, those beds seemed wonderful to us and we slept like logs. The covers were so deep we could get lost in them. If I hadn't awakened each day and dragged Oscar out we would have spent all day there.

We were awakened in the mornings about 8:30 by the Paris taxi horns. It is the great number of these buzz carts that keep the air of Paris well circulated. They tear around the whole burg but they seldom hit anything. They go so fast that one cannot even flag one for a ride except at street corners. The busses have practically no hoods and small motors, but they sure can go...

I had many interesting experiences on the Metro. I met one of my former college classmates who was going to Italy. He was sure surprised and so was I. It seemed so strange to be crowded in among so many foreign tongues. Often we met a man or woman who had lived in the United States and could speak English.

There are no elevated railways in Paris. The streetcar service is fair and cars are all operated by women or by men who are unfit for military service. It is so funny when one tries to tell one of these women where he wants to go. They answer with an explosion of French which tells you nothing and you are as far along as you were before asking. However, there is nearly always some one in the car who can speak English, or "Parley Anglais", as they say.

The subway is very good except that one has to always stand up as there are only a few seats. Each car has six double sliding doors, three sets on each side. They have first, second and third classes, there being a penny difference in price.

It is certainly hard to find a place by walking because the streets go by one name one block and by another in the next. They are like a huge web. They run in every direction. The Opera is the center of the whole town. I stayed about one mile from the Opera.

The whole city is pretty dark as there are only a few lights and they are on short posts.

The painted women of Paris are ever present and I believe are one of the features of the city. A fine old or rather middle-aged American doctor told us that his son is up in the front line and that he much prefers for him to be there and considers him safer than on leave in this city. Perhaps he exaggerated but he volunteered the information without being asked.

Letter #64

Laura to Erle from Paris
Hotel Edouard VII
Rue 8 Place Edouard VII
Paris
Adresse Telegraphique: EDOUARTEL-PARIS
Tuesday Feb. 12, 1918

My Dear: —

I almost sent you a picture post card but I feared to insult you so. The other girls have gone to a movie but I was too sleepy so staid home.

You didn't ever really think that I'd come to Paris, did you? I didn't but here I am. I shan't try to describe it to you because I can't and besides if I could you probably wouldn't be interested. I'll say this much, tho. It's by far the most beautiful city I've seen. Everything is so artistic and well balanced. There's no raw corners as we have in our own country. And London is not beautiful at all. It's dingy. You can ride for miles and it seems that on every corner or every second corner at least there's something with a story to it.

But it's a great handicap to speak no French. Conny has picked up quite a bit and manages to get along nicely so she is our business manager.

How you would enjoy it here! You'd fairly wallow in beautiful things. There are wonderful buildings with real architecture, wonderful cathedrals, great wide boulevards and everything balanced.

We came to one of the best hotels and we are surely enjoying the comfortable beds with the boxsprings and linen sheets and silk comforters. They are such a change from our huts. The French windows with the heavy brocaded curtains are like a story book and even the tiled bath room is a delight. I hadn't realized how crudely we've been living till I arrived here and saw real clothes again.

Last night we went to the "Follies". It was all right as far as the show was concerned but the behavior of part of the audience was unusual—at least it seemed so to us. If anyone behaved so at home they'd be put out of a respectable theater.

We have tickets for the opera for two nights. We are planning to see "Samson and Delilah" and "Mona Vanna". Eight months without good music or "shows" has left a big hole to be filled. . . .

Laura with the Anzac troops at Versailles in February 1918; the troops' curved side-brims hats mark their nationality.

Letter #65

Laura to Erle from Paris
Feb. 18, 1918

Dear Buddy –

I've just discovered that I didn't put sufficient postage on the other note I sent you from here so it probably won't reach you. Anyway I had very little of importance in the note.

It's a week tonight since we arrived here and we start back tomorrow morning. We've sort of hit the high places and have seen a good part of Paris and now I'm ready to go back to Camiers. I'd really like to live here a while just to have time to wander around and look but I can't so now I'm satisfied.

Saturday we spent at Versailles. I had no idea how beautiful the grounds were. The landscape gardening is perfect. I liked them much better than the palace altho that was wonderful. The trouble is that we've tried to gobble too much in a week and I'm all satiated with beauty of structure.

The atmosphere is so soft—so different from the glare and the cleanness we see in our own landscapes. You'd love it all.

This evening we went to the Café de Paris for dinner. It almost completed the destruction of our finances but it was worth it, the girls thot. I didn't like the atmosphere nearly as well as some other places I've seen. There was too much show for real comfort and too much of a glare. We went twice to a real French joint where no one spoke English and all the waiters were much interested in our appearance, strange language, etc. that I liked better and the food was better cooked and nearly as well served.

The painted women of Paris are ever present and I believe are one of the features of the city. A fine old or rather middle-aged American doctor told us that his son is up in the front line and that he much prefers for him to be there and considers him safer than on leave in this city. Perhaps he exaggerated but he volunteered the information without being asked.

So many Americans and Australians are on the streets all the time, I've been wondering if they were all on leave. But I suppose they are all getting in leaves before the Spring Push begins.

I do wish I could come on home now instead of back to camp. I've been half homesick all day.

The girls are planning to go to Rome on our next leave. It would be very nice but I'd rather we'd be sure of getting home to the States this Summer or Fall, wouldn't you? No? Perhaps after another few weeks in Camiers I'll be glad to plan anything. I hope we move soon back of our lines. I'm ready for that change anyway.

I'm going to have a lovely hot bath (pronounced bawth) and go to sleep.

I'm so glad that an English officer isn't going to censor this for me. I don't mind the French one. . . .

Letter 66

Laura to Erle
"Home"
Feb. 21, 1918

Dear Buddy: —

As you will note we are "home" from vacation or leave as we call it in the army. We did have such a pleasant time and saw many beautiful things. But I've told you about part of them all ready in the notes I sent from Paris. One note will probably not reach you because I didn't put enough postage on it but there was nothing of importance in it—just to let you know that I hadn't forgotten you even when in Paris.

I went to one of the huts this time but I hope I shan't be there long. I'm all ready to move now and I want to go. We have heard nothing further about moving but the leaves are all being crowded in so fast that something must be doing.

I can't tell you how I hated to get off the train at our station. I wanted to go straight along home. I had had enough of Paris in the week and wanted a little bit of Indiana and Chicago. Every day I have a new and fuller appreciation of our allies and the way they've stood the long strain of three and a half years. While in Paris we met a number of officers. I think best of all were a group of three Australians whom we met at the tomb of Napoleon. Then on the train returning we talked to a Canadian who holds his commission in the Imperial Army. Of course they were all chance acquaintances but we were with them long enough to get a pretty good idea of them and their attitude toward things in general. I'd be glad to know heaps more like them.

Of the Americans we met some were very pleasant. Almost all were quite diffident and not inclined to be friendly—you see them all over Paris looking at things, standing in the background with a sort of bored expression. Two young doctors we met I longed to spank. They were of the spread-eagles, Thank-God-we're-with-the-Americans type that have seen no real war service and have no conception of what the word means either to individuals or to nations. And yet they were pleasant boys—you couldn't call them men. The one was quite attracted by my ruddy locks and told me about his sweetheart at home in Mississippi. Evidently her hair resembles mine. But they are the kind that makes Americans disliked and distrusted by foreigners.

The Canadian (mentioned above) told us that he had been much surprised at this attitude of the American officers he had met. He said that instead of being noisy and friendly, as Americans are reputed to be, he could scarcely get acquainted with them at all. They seemed even more conservative than the English, he thot. . . .

Sat. Feb. 23

I'm still in the dark as to whether you'll be released from Unit 11 and get to come over for R.C. publicity work but I'm hoping and praying. But even if you don't succeed in putting that thru there's the big work of publicity at home and surely you've done more than many others in that alone.

But I do want you to come. It will mean such heaps to you and will be a very great experience. And then just imagine how delightful it will be if you can manage to penetrate as far as our unit. It makes my heart turn flip flops just to think of it. I foresee that I'll have to keep my hair curled continuously lest you catch me unawares. Oh, Kid!!!

You see my leave is all over and I'm not due another till four months are up. I wonder if by any possibility you are in France now. . . .

So you found a pleasant atmosphere in both Washington and New York. I've never been to the former but my one day in the latter city left an agreeable memory. I'd like to live there a while I think. . . .

I do so want the war to finish soon—funny desire—but a practically universal one. There's so much to do and so much to plan and everything and I wonder how these people have stood the pressure for so many long months. I'm afraid we Americans aren't very good stickers.

I do so want our troops to make good this spring. We have good stuff in us and I want to show these Europeans that we aren't all bluff. Our own papers are a little discouraging just now with their criticisms of the conduct of the administration but surely they will bring about some needed reforms. [***It is interesting to note from references in Laura's letters that, just as in more recent conflicts, the newspapers in 1917-18—the sole news media of the era—evidently criticized the president and government seemingly without restraint over the course of the war.***]

We get the Tribune fairly regularly now. In fact it comes thru better than our letters have recently. . . .

Pvt. Baker's memoir:

February 24

Fritz over. He was taking photographs so stayed up three or four miles and our guns did not come near him. Some duds dropped from our "Archies".

Letter #67

Laura to Erle
Wed. afternoon. [Feb. 27]

. . . I've been having a half day and a very pleasant time. First I went for a walk down to the beach then across to another road and by the little brook and picked Pussy Willows. Yes, sir, that's what I did! The bushes are beginning to show a little green and all the buds are swollen. And yet the spring is very, very slow. We've had practically the same kind of weather since Jan. 1st—cool, damp and misty with an occasional sunshiny day.

Instead of coming on to our hut for tea we went across to a club for nurses and "formed up" in order to be allowed to eat there and had tea and biscuits and black currant jam (doesn't that sound English?) and some delicious home made cakes. Then we explored the club house which used to be an artists home and found everything quite satisfactory. We'll probably make them glad with our presence rather frequently. It has a pleasant atmosphere and is a change from our other surroundings.

And yet it's the first time I've gone inside its doors during the whole nine months we've been here just because the spirit didn't move me to go in.

After tea I came home and took a nice hot bath and have decided not to go to supper so you can imagine me being just as lazy as can be except for an occasional poke at the fire.

I started to read "Mr. Britling sees it Thru" but haven't succeeded in getting wildly enthusiastic thus far. Perhaps I will be later but I doubt it. It's very hard for me to stick to anything so long as an H. G. Wells novel. I prefer magazines and newspapers. I very much enjoyed the articles you sent clipped from Atlantic Monthly. The one was "The Decline of the Berliner" by the Phillips and the other "France" by Galsworthy. They were both interesting and readable.

I've heard a rumor that there are twenty bags of American mail in tonight. I hope it's true and yet from the length of time the letters have taken recently to come I hardly expect anything more than a love letter or so from Miss Stodola. Yes, she still loves me and exasperates me most horribly. I very nearly swear when I see her writing on letters and packages. And still they come. Besides a letter each week she has sent at least four boxes of candy, a sweater knitted by herself and a large Xmas box of assorted foods and confections. And I still live on!!!

. . . We heard today that Besley is coming back to be our commanding officer. Evidently politics and not merit make for promotions. I hope he isn't representative of the men heading the other American units. But why rave? The war can't last always.

I smiled at the man passing out "green envelopes" today and he opened up his heart and gave me five. Luck? Well I should say so~! They forgot to give us any about half the time and I certainly don't want these goofs here to censor my mail.

We also heard that the British are bargaining for us for another year. If we don't move I'll be most horribly disappointed. I want to move from this neighborhood even if we're still with the B.E.F. instead of the A.E.F.

Today the end of the war seems so far away and indistinct. But I refuse to set a later date than 1918 for its finish. That's surely long enough for everybody.

Russia is surely making a mess of things. She is to be pitied but she has chosen such an inconvenient time to act up. Why not have her tantrums in peace times when what she did wouldn't specially bother the other peoples of the earth? . . .

The mist has turned into a gentle patter on the roof. It makes me long for so many things I haven't ———- and yet I've no cause to complain. We're comfortable here and certainly conditions could be far worse. . . .

Later

My curiosity is entirely satisfied as to the contents of the 20 bags. I didn't get anything and my roommate received two Tribunes. But I suspect that there were really 2 bags instead of 20. Mail reports are always greatly exaggerated. . . .

Pvt. Baker's memoir:

February 26

Fritz over again—about three miles high. Lots of noise from our antiaircraft. One dud came down through our kitchen roof and smashed a hole in roof and floor. As luck would have it, no one was hurt although there were lots of men around. We are growing accustomed to Fritz coming over at noon to take photographs every clear day. The airplane is truly the eyes of the army.

March 1

Cold, wet, raw day. Nothing doing. Regular routine of caring for our many heavily wounded cases and acting as undertakers for those poor devils who die in our wards. We get them ready for burial.

March 3

Fritz over at 12:00 noon. He was about two miles high and it was apparent that he was taking photographs. Photos can now be taken at an altitude of three or four miles with a telescopic lens. Our anti-aircraft guns (Archies) bang away tons of shells but rarely come close. One shell came up underneath him but he went up higher at once.

Spring 1918, March-May: German "Big Push" Threatens

Something in the Air
[Chicago Daily News, March 1918]

GAS MASK DRILL
FOR CHICAGO LADS
"The Canary" Puts the
Stretcher Bearers Through
the Paces

By Norman Klein
[Special Correspondence of the Daily News]

Chicago Base Hospital No. 12, Somewhere in France, March 2—"Take my advice, mates," said the Canary, with an admonitory shake of his bamboo stick; "take my advice and learn to put on your box respirator in six seconds. Fritz ain't got nowt to speak of when it comes to perlite society manners—'e won't give no warning. So be quick, quicker'n you ever was before—aye, quicker'n you'd be if the colonel came up an' asked for volunteers for the home guard."

The Canary was a Tommy instruction sergeant. He was called the canary because, like all the instructors back of the British lines on the western front, he wore a yellow brassard on his left arm. This sergeant was an old trench fighter; the four gold stripes on his sleeve were the official evidence of four wounds he received up the line. And so we thirty Americans grouped around him in the valley of these Picardy hills and listened with ears cocked.

There were twenty enlisted men among us, mostly Northwestern university undergrads; the rest were officers under Maj. Milton Mandel of Chicago and a few nurses. We had just been selected to form a reserve casualty clearing station squad and we were elated and excited at the prospects of going to the front.

Just now the squad was getting ready for the live gas, shrapnel and aerial bombs of the third line trenches. We were marched out to an antigas station, where we were to be "gassed" just to get used to the stuff...

Each of us had one of the box respirators newly issued by the British. This respirator consists of a rubberized face mask connected by a rubber hose with a red can of chemicals. The soldier, breathing through his mouth, inhales the purified air which comes through the chemicals and then via the rubber hose to his mouth...Glass goggles are in the mask so the soldier can see and fight with the mask on. A can of chemicals lasts twenty-four hours...

"Every British soldier must keep his gas mask at his side continually," says an order just out. Yes, and he must carry it on his stretcher when he is wounded, carry it from the emergency medical dugout in the trenches to the casualty clearing station, to the base hospital, and when he gets well carry it back to his battalion. Gas attacks are unexpected, and a soldier must be ever ready.

Diary of Dr. Neal:

Mar 3 1918—Russians declare peace with Germany.

Letter #68

Laura to Erle
March 4, 1918.

My dear Publicity Man

. . . I've been so lazy ever since I returned from Paris. We haven't been busy and our ward has only light cases so when the beds are made and the few dressings done and medicines given we have practically nothing to do all day. So I sit by the fire and listen. Just now we have an Irishman who is quite an entertainer. He is from Cork and has a proper brogue. He was in the Mons Retreat and has been out ever since. He has been recommended for the D.C.M. but hasn't received it yet.

Then there is an R.A.M.C. Sergeant (I'm always forgetting that you probably don't know what all these initials stand for—"Royal Army Medical Corps") who has a Military Medal with a bar. The bar means that he has won the medal again. The Military Medal is just a tiny bit of ribbon so [***drawing of strip with three vertical bars***] of navy blue with some tiny white and red stripes in the middle. The bar is a tiny silver rose ***[drawing of rosette]*** that is sewed on the middle of the ribbon. I haven't yet succeeded in finding out how he won either the medal or the bar. They usually blush and say they don't know or else change the subject.

He is a well educated chap of excellent principles and I'm very fond of him. As perhaps you'll note I'm terribly fond of these Tommies. They are a likeable lot.

Did I ever tell you of one Jimmy Nichols, a Scotch boy with frost bitten knees? Surely I've mentioned him some time. He went back to the base this last week and it almost broke my heart. If I had had a younger brother I'd have liked one just like him. (People always select their relatives, I notice) His mother sent Conny and I each a tiny plaid bound volume of Scot's Lady of the Lake. Their home isn't far from the scene of the lake.

Today we had a coffee party in our room. We bought a loaf of bread at the canteen and had quite a feast and drank cups and cups of the good coffee. . . .

I hear now that we aren't going to move—but neither is that authentic. However I understand that the British cannot furnish a staff for the hospital so an American staff will be kept here. Therefore it is a probability that we'll remain here. And I want to move so badly. We've been here nine months and of course being myself I'm ready for a change—fickle person that I am! . . .

Pvt. Baker's memoir:

Sunday, March 10

Fine day. Spent it on duty. Am all alone in ward as the other man is in quarantine. All heavy surgical cases. I have many bad cases and all the worst kind—wounds in head, chest and stomach. Many have open wounds that are draining so they can heal up clean. It is the wreck of war.

March 11

Fritz over. Six batteries of antiaircraft guns fired at it. Several shells came down a z-z-z~ing—duds. Fine day.

The Sweethearts Disconnect

Letter #69

Laura to Erle

This Active Service envelope, sealed by censor #6079 and addressed to Erle at the Western Newspaper Union, was marked on the reverse, "Rec'd April 14 1918 at Camp Dodge." It had been forwarded by the office to his new U.S. Army address.

France

March 11, 1918

My dearest Buddy:—

. . . Before you'll get this you'll have crossed the Atlantic. Perhaps I'd be wise to wait and sent it on to the R.C. headquarters in Paris. But if you were held up any place it's be so long before it reached you then. So I'll be good and send it on to the W.N.U. [***At this point Laura still believed he was coming to France with the American Red Cross as a publicity man rather than with Base Hospital 11***] . . .

Letter #70

Laura to Erle

March 13, 1918

[No salutation]

. . . Did you know you have a rival? He is Cpl. Pattison of 10th Rifle Brigade. He has a sprained knee, a pimply face, a tenor voice, the mind of a 14 yr. old and he sings or tells jokes (so-called) continuously and is always in trouble over his messed up bedside table. And look at the beautiful poem he has written for me and to me! I'm sending the original home as I thot you'd like a copy just as well.

I don't remember anyone writing poetry to me before and I'm quite overcome. . . .

I had a day off yesterday and went down to [censored] to see the cathedral. Had only a fair time. Perhaps you'll be here by the time my next day is due and we can wander around over the hills and to the neighboring villages and have a real visit with each other.

I heard today that we are to stay here. I've stopped worrying about that. All I want now is a real victory over Fritz so that we may all go about our business away from this tragedy. . . .

Today I had a beautiful package containing two boxes of delicious stick candy and I felt like a small child I was so pleased. Didn't you laugh when you received my request for it? I was quite ashamed of myself afterwards for asking you but that was too late after I'd mailed the letter. You are a dear and [I] do thank you most heartily for the candy and the much coffee that you've sent and all the trouble you've taken with the packages.

Also I received two pkges of papers today—the first for ages. The two were mailed a month apart but came to me at the same time. . . .

A TESTIMONY TO SISTER HUCKLEBERRY

Whilst in the 18th General
We were treated, not too bad,
But the only one we care for well
Is the Sister that keeps us glad.

She is little, but very kindhearted
To everyone in bed.
She's got a smile, that's never parted,
If parted, we would be "dead."

We wait until the morning's come
That we may see her face
With that sweet smile, that's never gone
And such a charming grace.

Her duty done, then off she goes
To bed, we think 'twill be.
But when she's here, for all the day,
She's like a busy bee.

We do not know, how it would be
If she left us, to part.
We're sure the "Boys" they all would say
That it would break their heart.

May God bless her, and keep her well
For ever more to be.
In the ward, she's needed, so very well,
She gets a Praise from me.

With the Best of Wishes
from

The patients in Beds No. 1, 2, 3,
21, 22, 23 & 24
D Ward 18th General Hospital

France

(Attributed to Cpl. Pattison, 10th Rifle Brigade, B.E.F.)

Pvt. Baker's memoir:

March 14

Over half of the unit, Colonel Collins and Major Mandel were out working on the ball diamond. Colonel Collins wants the boys to enjoy good, clean sport and so is encouraging us. Major Mandel is a fine old doctor.

Letter #71

Laura to Erle

[Active Service envelope, censor #3961,
"Rec'd April 14—1918 at Camp Dodge]
Friday Evening
March 15, 1918

Dearest Buddy: —

. . . How I wish I knew if you've started for France!! It's most a week since I heard that you'd had orders to start and that is a month ago today. Surely you are on your way by this time.

If you get an opportunity to come up here to our camp I'll try to get a day off to tramp around the country with you. But I'm trying not to count on it too much lest by some accident we shouldn't meet. There's so many ifs in this big war you know. Sometimes I think there's more ifs than anything else. Now I wish it were all over!! There's a selfish reason back of it all but aside from that it is all so unbalancing. And as yet we can see no end to it. It will take some good hard fighting and yet our country isn't ready, is it?

It's quite discouraging to read our papers and yet I know that they have had the same troubles in the other countries at war. We aren't all angels yet. But I'm almost me when I munch this good stick candy. I smile like a cherub. Yes, I think if nothing else would succeed in holding my affections stick candy could, the Saturday Evening Post and all the magazine sections of the Sunday papers together with the various long typewritten letters should do the trick. . . .

. . . I went over to the mess and danced a couple of dances to the victrola music and then came back to our hut. I get very tired of the concerts—they're so much alike and the dances are so polky with always the same people. When the war is over I don't think I'll even want to see any of these people again. I get so tired of them all. . . .

. . . The war can't last forever, can it? I do wish this year would see its end. The Kaiser surely did stir up things when he started! I'll bet he wishes he hadn't. But just now that is of no particular benefit to us.

We are gradually getting busier in the hospital so I came home yesterday and today [am] quite tired out. . . .

. . . I went over to the mess and danced a couple of dances to the victrola music and then came back to our hut. I get very tired of the concerts—they're so much alike and the dances are so polky with always the same people. When the war is over I don't think I'll even want to see any of these people again. I get so tired of them all. . . .

Pvt. Baker's memoir:

March 16

We are easing up a bit as there seems to be a lull in the fighting. We expect big doings very soon.

Took hike with one of our good nurses in the evening. All the nurses, men and officers off duty take to the hills for recreation.

March 21

The great Hun drive on a sixty mile front, south of Lens, began. Their objective is Amiens. We have orders to evacuate on a moment's notice, but to put all wounded on board ship first. The English have a lot of ships in the Channel in readiness for any emergency. Germans are determined to break through to the Channel and separate the British and French.

We are swamped with wounded and the poor devils sure are pessimistic. They say they never saw so many Germans in three years. The Huns are well armed and outnumber the British greatly. It sure looks bad for us. Fritz still has the supremacy of the air. The French are rushing a division up north to help our men.

Letter #72

Laura to Erle
France
March 21, 1918

[*No salutation*]

Yesterday, m'dear, arrived your letter of Feb. 27th. I'm so very, very glad to hear that you've received a cheerful letter from me and that you are glad at me again. And I'm so sorry to hear of the mixup about the orders and your dilemma as to the Red Cross assignment and the mobilizing of Unit 11. By this time it's all settled, of course, and it will be at least another two weeks and possibly a month before I'll know what has happened to you. I shan't say which I hope it has been because you already know my feelings on the subject. This long distance correspondence is very unsatisfactory, isn't it? . . .

We're quite busy in the wards now-a-days but nothing like when the push is on. I'm beginning to hope I'll go to a heavy surgery ward by the time the rush comes. They're heaps more interesting and have heaps more of work.

There's almost no news. Everything moves along in the same way. The only thing existing that ever happens is a mail coming in just when we've given up hope. . . .

I feel so ashamed of myself for being so "fed up" after ten months when I think of the way our Allies have stood up under three years and a half of much worse trouble than we are having. Please forgive the whine. I'm not really downhearted. I'm just making conversation.

Brother Silas is home now and convalescing nicely even tho slowly. He was pretty badly injured. My last letter from home came in fifteen days. Wasn't that quick?

German Spring Offensive Nears

Pvt. Baker's memoir:

March 22

One of the greatest battles of the war is in progress. The Germans have captured over 16,000 British, 200 guns and millions of dollars worth of equipment. We still have hopes that the Lord is with us and we feel confident that something will happen which will check this terrible offensive. Rumors are many, and wild. The wounded tell us that the only hope is the Yanks. When will the Yanks get into the fighting in real strength? The Germans, nearly 500,000 of them, are pressing on and they expect to come on through to the Channel. If they do, England is through in this war. We are all with our backs to the wall. We have only thirty miles in which to maneuver a great army, guns and our wounded. The English are a little "down" but not "out". Lloyd George is issuing strong appeals to the British people to "keep their tails up", etc. French peasants are really afraid the Huns will come through this time and they will all be prisoners. Thousands of civilians are leaving their homes around here and taking to the open road for the South to escape being captured. We are overwhelmed with thousands of wounded.

March 23

The terrible battle is still on. We are very busy day and night. There is not much sleep for anyone. Received seven to nine hundred wounded in twenty-four hours. Worked many hours. Huns have captured thousands of English and hundreds of guns and tanks. Our only hope now is in what Uncle Sam can do. Every one is asking, "How many Yanks in France?". One French official in Paris published the facts to help keep up the spirits of the Allies. He said the Yanks were ing in at the rate of over 500 an hour, or 12,000 every twenty-four hours, and soon they would comprise an army large enough to take the offensive by themselves. I have the utmost confidence that we will win the "guerre" yet, but I am one of the very few optimists around here. We have evacuated every wounded man possible to England.

Diary of Dr. Neal:

Mar 23—"Big Bertha" 70 miles distant bombarded Paris for 7 hours with 9 1/2 inch shells.

Pvt. Baker's memoir:

March 24

We hear the guns every night and see the flashes, veri-lights, etc., at the line. A tremendous, continuous, thunderous roar. We have air raid signals every clear night.

Letter #73

Laura to Erle
Sunday Evening
March 24, 1918.

Considering the action and threats of the German Push recorded by George Baker, Laura's next letter is the first even to hint at the anxieties she and the other nurses must have been feeling. Likewise, her sweetheart Erle—and her family in North Vernon—must have been worried about the possibility of a German breakthrough at nearby Amiens, as reported and warned in American newspapers in great detail. Would she be injured or captured? No one could predict.

Dearest Buddy:

Tonight I heard there were twenty three bags of American mail in and I went over to the mess in high hopes. I received a slushy mash note from Miss Stodola replying to my effort to squelch her in the only letter I've written her since last June. Honestly she drives me distracted at times. I hope I'll get some other mail later in the evening. They surely haven't had time to sort all that so soon. Recently most of our big mails have reached us on Sunday—but you aren't interested in such things at all.

Of course you know that we are very busy just now and expect to be still busier. Quite a number of our nurses are off duty sick and not all have returned from vacation so we are a little short handed. Probably most of my communications from now on will be along this line but then you know there isn't much variation in the topics I can speak of.

Conny is off duty sick and no one has been assigned to the ward to assist me; we have been cursed with the most worthless orderlies in camp; tomorrow is inspection day; the ward is dirty; we've lost all but one broom and that one has most of the bristles worn off; I don't like our medical officer; and Goodness only knows what other troubles I have. Now isn't it an awful thing to let such things bother me when this huge push has started? Really I'm ashamed of myself. If only Conny were there to run the place I'd not mind anything I'm so tired tonight. . . .

Tuesday 1 AM

Dearest: -

I'm again on night duty and I'm not quite busy enough tonight to not be a trifle lonely. We are running a fairly heavy hospital for the last few days and I was glad to get on nights to get away from the worry of the day time. You see, at night I have only my work to look after while on days its not only mine but everybody else's also.

Gee! It's taken me most half an hour to write this much after making the statement that I wasn't busy. But you see everybody started coughing—I'm in the chest ward this time—and then the perfectly good natured snail-like orderly came in and talked and so I'm not progressing very rapidly. . . .

Today I've heard so many rumors of all sorts about activities but none are authentic. Anyway I couldn't talk about them on paper so I'll just have to wait. Always the most interesting things are tabooed by the censor.

Probably you know thru dispatches more than I do thru rumor so it is unnecessary to tell them over. . . .

Pvt. Baker's memoir:

March 25

Very busy. Every able-bodied man is helping move the huge army of gassed and wounded men. Long, weary hours are spent, day and night, carrying the wounded from ambulances to the hospital. Troops are very gloomy and are fearful lest the Huns push on and break the line. There are terrible cases in. They say the Huns are determined on this last stand before America gets many troops over. It is their only chance. We are all praying that the U. S. will have an army by itself soon. We hear there are over a million Yanks in France.

March 26

Great battle still on. Troops, fresh from England, are coming through day and night, and rushing up to stop the great Hun offensive. Terribly busy. We have had as high as seven to eight hundred wounded men arrive at the hospital in twenty-four hours. This means many weary hours of stretcher bearing. There are about 25,000 wounded in this area now. I am about all in, but the sight of these wounded men spurs me on and I keep going. I have about thirty-five in my ward to take care of.

The Hun aviators know every inch of ground between the Channel and line and they bomb every place they think the British might have ammunition stored, or might have troops, etc. One of the largest ammunition dumps is located just a thousand yards away from us, down in the sand dunes and pine trees, and it seems as though the Fritz tries to get it every other night. If he ever gets a direct hit on it, as he did at one near Amiens, we will all meet in heaven. There is enough stuff in this dump to blow up the north half of France. We watch the searchlights as they try to pick the Hun flyers out of the black night. When the Huns fly over these lights they drop 20-pound bombs. Believe me our men at the search lights and batteries of sky shooting guns (Archies) sure need the protection their deep concrete dugouts afford them.

The Huns are trying to break our morale with these raids and with their success at the front, does look as though they are on top. They have now reached the Somme River, twelve miles east of Amiens.

One company of American Railway Engineers stopped building a railroad here and, taking up rifles and grenades from the dead British, dug in and stopped a wide gap of a mile or more in the line. They fought desperately and held on until reserves were rushed up to them. They all received decorations for this feat.

Nearly everyone is getting his "wind up" which is the English expression for being scared. We are all scared but no one likes to admit it. The English wounded say there must be a couple million Huns turned loose in this sector. They can't imagine where all the Huns come from. *[German forces had been transferred from the Eastern Front with Russia to the Western Front in Belgium and France following the October 17 Russian Revolution and subsequent peace agreement between Germany and Russia.]*

March 29

There are troops, troops, everywhere. Many are England's seventeen and eighteen year old boys. Their country is at stake.

I held the hand of one poor fellow and helped him write his last words to his wife. He died an hour or so after. I have forgotten his exact words but they were substantially as follows: "Dear Wife: Am expecting to go West soon and wish to say Good-bye. Your loving husband." The next day his widow arrived and I told her about him and of his last thoughts. I don't know of anything more trying on a fellow than to see these poor faithful relatives from England at the deathbeds of the ones who have given their all to save their country. Whenever it is possible the relatives are sent for, but often they come too late.

March 30

Have been helping some of the poor fellows write to home-folks.

Have had my first experience at embalming, or getting a corpse ready for burial. We used to have wooden boxes but there are so many dying now that it becomes necessary to bury the poor devils wrapped in army blankets only. The percentage of deaths, however, is very small in comparison with the vast number coming in for operations. Many of the otherwise unscratched men come in with terrible "trench feet". In many we cannot restore any life or circulation at all and this means the loss of one, and sometimes both, feet. "Trench feet" is a peculiar and awful type of affliction. The feet turn purple and then black—just die. "War is Hell", as Sherman said.

I held the hand of one poor fellow and helped him write his last words to his wife. He died an hour or so after. I have forgotten his exact words but they were substantially as follows: "Dear Wife: Am expecting to go West soon and wish to say Good-bye. Your loving husband." The next day his widow arrived and I told her about him and of his last thoughts. I don't know of anything more trying on a fellow than to see these poor faithful relatives from England at the deathbeds of the ones who have given their all to save their country.

Sunday, March 31

We are praying the great drive will let up, but the Huns are pressing for victory and spending their all in trying to come through and capture the Channel ports. We happen to be right in their line, so can only hope for the best. There are all sorts of rumors.

The Huns continue to raid our area about every other night so we have no dearth of excitement around here.

We have three nurses, two light duty men, myself and the doctor as the Ward Staff. Besides this, it keeps four or five "up-patients" busy running for the hundred and one things that the patients require. This gives some idea of the care given and the amount of work entailed in feeding, changing bandages, etc., etc. Some of the patients have no use of their hands at all. Others cannot raise their heads. We have men wounded in every conceivable and horrible manner imaginable.

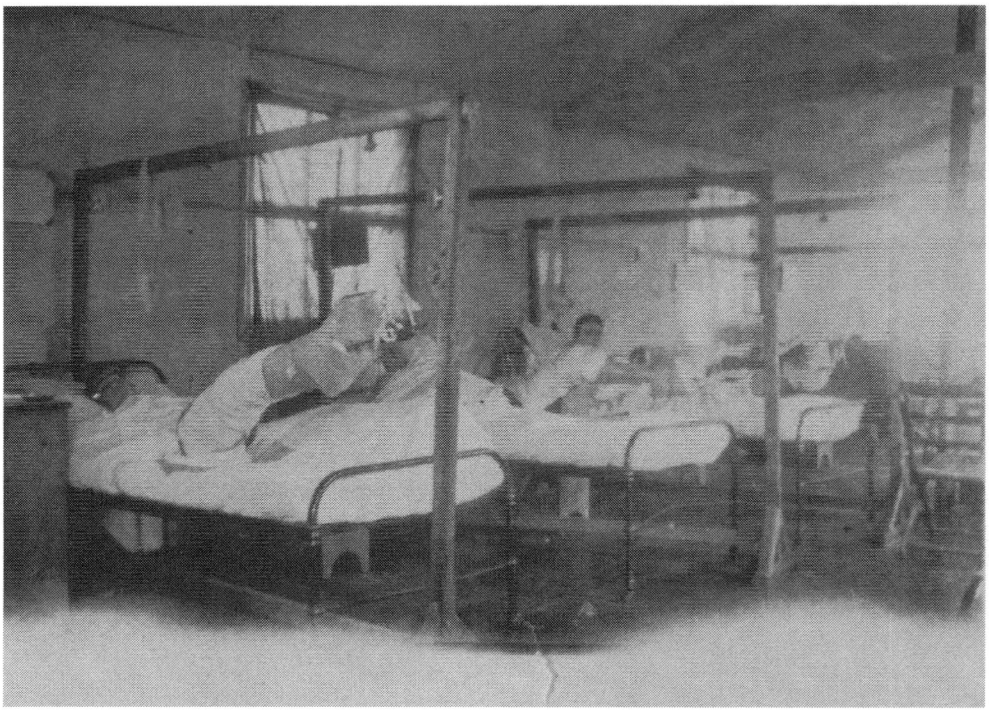

Fracture frames with wounded soldiers.

Letter #74

Laura to Erle
France
12:05 AM. April 1st 1918

Dearest Buddy: -

Easter is just past. It has been a beautiful day and I've slept almost all of it away except a bare two hours of daylight when I took my usual little walk toward the sea.

Do you remember Easter Sunday two years ago? I went with you down to the Holy Name Cathedral and you had to give me your prayer book to keep me quiet during the service. There was much pomp and ceremony and a beautiful white altar with heaps of white lilies. . . .

I had a queer sensation a day or so ago. I was in Bess' room lying on her bed reading when my attention was caught by some one in the next room reading aloud that Red Cross article you wrote from my letters. It had probably been sent as a clipping to the girl reading it. Then I heard the comments.

"She's evidently in a British Hospital."

"But those red comforts weren't sent for the wards. They are our own."

"Somebody evidently stretched her imagination and was writing for effect."

Well, perhaps I was but it was all so real to me that day I wrote you. And you had evidently misinterpreted the red comfort part of my statement ***[Letter #54, January 1, 1918]*** but that didn't alter the yarn's value as reading material. That last comment was Mac's. Knowing her as I do I didn't feel hurt but rather amused. It's all in a life time, isn't it? Now do you feel evened up for "that impersonal stuff"?

My ward is so quiet tonight and I have an orderly to help me so I think we'll get along nicely. In fact if you won't tell anyone else so it will get back to the authorities I'll tell you that I consider myself fortunate to have but one ward and only thirty patients. It's what I might call a snap for these busy times. Most of the cases are chest cases—bullets or shrapnel in the lungs—and there is a strict ruling prohibiting their removal to England before three weeks have passed from date of wound. That means that we do not evacuate many nor get so many new ones by convoy as the ordinary surgical wards do. But they are sick and uncomfortable. They sit up and puff and cough and can't get their breath and occasionally have an internal hemorrhage.

We don't get a great deal of authentic news here. It's principally all rumor. You probably know more about the real course of events than we do.

I wish I knew where you would be when you get this or while I'm writing it. I like to imagine you either at your desk or in your room while I'm talking to you on paper. Foolish idea, isn't it? Perhaps you are at Fort Dodge, perhaps you are on your way across, perhaps you are still in Chicago.

Gee! How I wish it were all over and I could really talk to and see you! Ten and a half months seems a long, long time to me. . . .

Pvt. Baker's memoir:

April 1

Very busy. Steady stream of wounded coming in all the time. British army have their backs to the wall and the whole army is very gloomy as their prospects for victory are poor. Every thought is on America. Can she get troops here for action? Many French have given up the thought of winning. The British Government is issuing messages of courage to the troops. Lloyd George, the well known war Premier, tells the boys to "Keep your tails up", or in America we would say, "Keep up your courage".

April 2

Very busy. Offensive still on and many wild rumors are afloat, such as the Huns having captured Amiens. Shells are falling on the cathedral. We get bombed nearly every night or are close to where bombs are dropped. We have many more guns and searchlights now. One Hun rode straight down a searchlight beam and opened fire with his machine gun and then dropped a bomb near it. The crew had all dropped into a deep dugout and were saved. We often watch the searchlights pick up the night raiders. It is a wonderful sight to see a plane with a full beam on it. They dive, zoom, spiral, etc., as they try to get out of the powerful light, and they are usually successful. Sometimes two or three searchlights get a plane in a pocket, then the guns all bang away at him. He either comes down and surrenders or is hit with a shell and killed in the fall.

Letter #75

Laura to Erle
[Active Service, censor #3150, still addressed to W.N.U. Chicago, "Rec'd at Camp Dodge May 6 1918"]
France
April 4, 1918

Dearest "Young Man"

. . . Tonight promises to be quiet but one can never tell at the beginning how things will be later on. I've just had a transfer from another ward and have him nicely settled. . . .

We have some Americans in the hospital now. They are engineers who were with the British. I haven't any in this ward but they are scattered over the rest of the huts. They tell some rather wild tales and their morale seems to be very good. In fact it seems to me that the morale of all the men we get now is decidedly better than it has been all winter. The German advance has awakened everybody up again.

Of course you know much more of the late news than we do as even the papers are irregular and when we are on nights we are apt to miss the paper boy. Then, too, we usually get a Daily Mail and I suspect the English Press of being rather severely censored. Of course, I don't know. I'm merely guessing.

Beesley is back and it has a depressing effect on the whole unit. He has certainly made himself about as much disliked and despised as any one I've ever known. We were getting on quite well without him. But I shan't bother you with unit gossip.

The night sergeant just came in and announced a convoy. I'll have to run out and turn down a bed or two.

April 7, 2 AM

. . . I've discovered that individuals count for little in a war but as long as you are alive and well, there is much to be thankful for. There's so many tragedies—physical and mental. I'll tell you some of them some day but of course you read them or similar incidents in most any war articles. There've been two cases since I've been on nights that sort of knocked me out for a while but one gets hardened to tragedies as well as to other things.

I shall keep on addressing your mail to Western Newspaper Union till I get a new address from you. Is that all right? If not it will be done before you can prevent it. . . .

The boy that is my orderly is the nicest youngster. He is an only child, and probably a spoiled one but he is making the most of himself and doing excellent work. Then I like to have him about. The last one I had was a farmer boy who reminded me of Wm. Hodge but this one does far better work. For a long time I had met up with only good-for-nothing men till I hated the word orderly. There is such a difference. It is disagreeable work but when a man rises above that part and simply does his part I feel like taking off my hat to him. After all it isn't the work itself but something inside the man that does it. Am I not right? . . .

. . .

I stopped there to write my report and take it up to the night supervisor. I've an awful ward tonight. One man is so delirious that he keeps everyone else awake. But that isn't what I started to tell you.

I walked back down the road with a cheerful sort of person. Said I, "I wish this thing were all over." Said she, "Why didn't you marry that man before you started?" "Oh, I'll do that when I go back." "Don't be too sure, young lady. I kidded myself for four years about a man and since I've been here he's married another girl." "It's a gay life and a funny world," I sympathized, "Sometimes I think it is either gay nor funny but what's the use? We might as well be cheerful."

Of course, she knew nor knows nothing about you nor your existence. That was just talk but she certainly meant what she said about her love affair and I had always thot her one of the fluffiest care free people in the unit. We certainly don't know much about each other after all. I don't know why she told me, I'm sure. But people do blurt out things of the sort sometimes without really intending to. . .

I walked back down the road with a cheerful sort of person. Said I, "I wish this thing were all over." Said she, "Why didn't you marry that man before you started?" "Oh, I'll do that when I go back." "Don't be too sure, young lady. I kidded myself for four years about a man and since I've been here he's married another girl." "It's a gay life and a funny world," I sympathized, "Sometimes I think it is either gay nor funny but what's the use? We might as well be cheerful."

Pvt. Baker's memoir:

April 4

Very busy unloading and loading ambulances and carrying the wounded on stretchers.

We have a lot of Chinese coolies around here for road work. They have strong backs and they live like cattle. They are kept in wire net pens at night like prisoners.

We have a German prisoner camp near us. The Hun planes dropped a couple of bombs near their own men and wounded a few of them.

The Chinese bitterly hate the Huns since a few bombs were dropped on the Chinese camps and a few of them were wounded and killed. They tried to storm the German prisoner camp and if they had not been prevented by the guards, they would have killed the German prisoners.

April 5

Busy, night and day. We have wounded everywhere. The great Hun drive up east of us at Armentieres started yesterday. Terrible loss of life to both sides.

Sunday, April 7

Went to church in the evening for a half hour. Very busy with heavily wounded men. Everybody is about crazy from handling wounded soldiers. It looks like old Kaiser Bill is going to have a vacation in London soon. We hear wild stories about the terrible Zeppelin raids on poor London. We can often hear the coast guns bang away, driving off Hun raiders.

April 9

Busy with heavily wounded.

April 11

Busy. Very bad cases. Laid out two or three men from my ward who went in a week. Have one terribly bad case now who requires more little cares in one hour than a dozen ordinary men. He is an Australian and is sure shot up badly. I think he has a dozen wounds. Some are shrapnel buried in deeply. It is certainly amazing the pain and grief these poor fellows take 14,000 miles away from home.

Letter #76

Laura to Erle
France
April 11, 1918.

Dearest Buddy:–

I wonder if you are reconciled to your big disappointment and how things are going with you and where you are and what you are doing and about a thousand other wonders. For all I know you may be yet in Chicago or perhaps cantonment or else started for France. You know someway or other I have a hunch that some day you'll be glad that things turned out as they have. Really we can see such a little way ahead. I do hope you've been able to readjust yourself to circumstances and that the unit has been called to active service. You'll be so much happier than just waiting. You've had such a year of waiting. . . .

I wish I could be as sure of when the war would end as I am of the durability of our mutual affection. Now that's quite a statement for me to make, isn't it? But please remember that when you get a foolish sounding statement and think I'm doubting you.

I couldn't have written you such things eight or ten months ago even if it weren't for the blooming ***[note her British slang]*** censor. But I'm getting so accustomed to him now that it isn't so bad. As long as I can get "green" envelopes I feel safe in saying almost anything. You can't know the feeling till you become accustomed to your shortage of "blue" ones. I suppose there'll be a shortage of blue ones as there usually is of green.

Of course we are busy but tonight is so much calmer than any we've had for a week that I'm taking a little time to visit with the orderly. He lives just forty three miles from Kansas City, is nineteen years old, is blonde and a very likable youngster.

The weather is misty and has been so for several days. But I don't mind at all when I'm sleeping days. I'm surely sleeping well—really better than I had at night for some time. I usually am asleep by 9 AM. Sleep soundly till 4 P.M. or even 5, then rise and make coffee and take a walk, eat supper and come to work. That is surely uneventful living. But then I didn't come over to live but to work and there is usually plenty of that.

Why congratulate me on the Red Cross article you compounded from my letters? That was your story. I'm glad it was so popular. It didn't look so badly in print, did it? Even tho written for effect. . . .

Letter #77

Laura to Erle
France
April 14, 1918

Dearest Buddy Boy: —

. . . The men have a song all about the different hardships and irritations they must put up with. The last line is "Oh! Oh! It's a lovely war." Then so often one of them will remark, "It's the best war I've been to." They are such fitting expressions for the feelings we feel. The war certainly upsets every one's plans for themselves most promiscuously. But never mind. After the Germans are licked we'll try to straighten things out for ourselves. Theres considerable fire works to be gone thru first but I hope it won't be a great while. You see I continue to hope and you must help me hope.

Yesterday was quite chilly and today there is a raw wind rattling all the windows and doors and making us shiver. It feels like last December. Again I'll rise and remark that I'm glad I'm having a nice comfortable little hut instead of two or three tents.

I was going to write you last night but one of the men was so delirious and restless that I hardly had an opportunity to read the headlines in the "Daily Mail." Any way there is so little to write about. I've told you about my men and the ward and the orderly and that all I do is eat and work.

Oh yes! I've thot of something to tell you. But perhaps I told you before that I had a photo taken for you while in Paris. They have just come and I'll be sending them along to you soon. There were two and considerable discussion prevailed as to which was the better. I'll let you decide for yourself.

I like the one with the stray lock of hair because the line of the neck is good and I like the pose. I can't say much as to the likeness because I never like my own pictures. I thot perhaps you'd like one and so that's a souvenir from Paris for you.

Frances had promised a photo to McKosh and Conny & I went along with her to have it taken. Then we decided to be frivolous also. I think Conny got the best of the bunch but most every one likes mine better. Frances got a very poor one and she is so disappointed. She did so want a handsome picture.

It's so funny to watch her write a letter to Mac as she calls him. We all have to keep still while she composes it. She writes a page, tears it up, rewrites it, tears it up, etc. till finally one is satisfactory. It's a fortunate thing for you that I'm not so particular else you'd get but few letters from me.

Now you know that it lacks just one month and two days of being a year since I left Chicago? It's a long, long time anyway whether you know it or not. But I'm sure you do. . . .

It's really strange how many of those pleasantest memories you seem to be connected with even way back. Do you suppose that has any connection with the present state of my mind & heart? I think so. There just never seemed to be anyone else that fitted in exactly.

Honey, every one is moaning so tonight even when asleep that I just can't write anymore. . . .

Eleven months today since I saw you. It surely seems like an age. I hope it won't be that much longer. There's so many things to enjoy in this world it's too bad that this world war had to come along and break up everything for everybody. "Oh Gee! How I do hate the d—— Kaiser!" Please note that that is a quotation.

The War Wears On

Letter #78

Laura to Erle
France
April 16 1918

Dearest Buddy

Such a perfectly delightful walk I had this morning!! Everything was covered with mist and drops of moisture. I meant to walk up on the hill and then go to bed but I couldn't come in so early. The fields were so green and the birds were twittering down below me and clouds of mist blowing in my face so I walked down the hill into the village, stopped to look into the church (why are Catholic churches so peaceful—that is if they don't clash?) and on to the little lake. Instead of coming on thru the village I walked on around the lake under the pine trees and then past the little brook that always looks like spring. The walk along the brook is called Lovers Lane but I didn't see any lovers. In fact it was too early for any one to be out except me.

There is a shrub of some sort that has blossoms like cherry trees have. They grow quite thickly along the road sides and past the brook so you can imagine how fragrant the air was.

I came home quite enthused but Miss C. refused to appear interested. I think she had meant to go with me but I was feeling exclusive and walked off alone. Is that snobbish or merely selfish? But, my dear, I get so tired of her and she is so perfectly nice and ladylike and intelligent and everything that I should be but am not. . . .

Eleven months today since I saw you. It surely seems like an age. I hope it won't be that much longer. There's so many things to enjoy in this world it's too bad that this world war had to come along and break up everything for everybody. "Oh Gee! How I do hate the d—— Kaiser!" Please note that that is a quotation.

I see by the papers that we are to be allowed no packages any more. That's too bad, isn't it. But we've had lots of them up to date. I yet have two lbs. of coffee & my room mate has three so we'll have enough for a while & then we'll subsist nicely on issue—tea. . . .

Nothing has happened today except an effort to sleep and a short walk toward the beach. I haven't been to the beach for quite a while. I think I'm too lazy.

I have just a week more on night duty. I'll be glad to come off even tho I've enjoyed it. I'll be delighted to go to bed at night.

The day orderly is on guard tonight. He just stopped in to talk awhile. He's an awfully nice boy also. Some of them are just as nice as can be in spite of all I've said about the ones that aren't nice.

A bunch of them gave a minstrel show this week and it was very good indeed. I did enjoy some of their songs. There are some excellent voices here. I'll stop now and run over with my reports to the night bunk [office].

Pvt. Baker's Memoir:

April 21

Went to church. Sending my patients to Blighty—have only a few left. Getting ready for big doings. On April 17th the U.S. Division went into the line at Montdidier. The 26th Yank Division repulsed the Germans at Seicheprey with heavy losses to the Huns.

April 22

Only five bed patients at present but two are very bad and probably will never recover. We cleared out the hospital by sending all moveable patients to England. We are getting ready for final struggle and things sure look blue. Everyone is living on his nerve and on his faith in God that the Allies will somehow come out victorious. Many of the nurses have had to go to rest camps because of nervous collapses, overwork, etc.

Letter #79

Laura to Erle
France
April 23 1918. [Tuesday]

Dear Buddy:

. . . It's over a month since Fritz started to push and yet nothing very decisive has occurred. I'll bet he's disappointed.

We've had a busy month here but have pretty well cleared out the hospital now evidently awaiting something else in the way of a push. Perhaps it's a counter attack.

I've just found a Ring Lardner article that I clipped from a Tribune a month ago and meant to send it to you. You've probably read it before.

Nothing much has happened during the past week. I'm all right—that is, I eat, sleep and work in the usual way. I can't really say that I'm happy but I manage to scrape along.

Tonight I sent out eight of my chest cases to Blighty and seven last night so I have a half empty ward. All that are left are certainly sick.

Thursday Evening

It's taken me two days to get so far and even now I don't know what to say. I want to talk but apparently there is nothing to talk about. I've been on night duty so long—thirty-two nights in all and I've grown cross and crabbie and you wouldn't like me at all if you were here.

Tomorrow I come off nights and a big reformation is coming off also. I'm going to try to be agreeable and do no grousing. And then I'll expect a little pair of pin feathery wings to push out thru my shoulders. Do you suppose the feathers will be red or bronze? . . .

So many things can happen in the course of a year or two.

I've planned to go over and see the little gray haired English artist lady Saturday

I read in the paper yesterday of a Spanish beggar woman (very beautiful) who had prophesied that the war would be over in six months. I've elected to believe her. I always try to believe the things I want to. It's much more satisfactory than the other way about and works out just about as well one way as the other. . . .

afternoon. I've never been back since last summer. I've intended to go but have always lacked the time or the money (I don't dare go a place like that without money) or else some one to go with. I've accumulated quite a little bunch of etchings, but I think hers are by far the best I've seen. . . .

Spring time never was meant for wars, was it? . . .

Letter #80

Laura to Erle
France
April 27, 1918. [Saturday]

Dearest Buddy:—

Your letter of April 4th from Camp Dodge arrived late last night. Also the one of March 27th and the Mencken review mailed on the same day. I'm glad that at least you've been sent to cantonment and I hope for your sake that you won't tarry long on that side of the water. You'll feel so much better about things when you come across here.

I wonder if you'll think me crazy if I say that I'm glad and I know that some day you'll be glad that you are called on to serve in the army rather than simply in the Red Cross! There's an element of personal sacrifice. . . . in this that you'd never have had if you had come across as R.C. representative. I know that the latter would have been much, much pleasanter and more in your line and you'd probably accomplish more actual valuable work that way. But I wonder if always there wouldn't have been a little doubt in your own mind. And yet I know exactly what your present "job" is and I know all it's discomforts and disagreeableness and every thing. . . .

. . . I've had a delightful day. I'm off night duty and I've put in the day shopping and walking and shopping. Then this afternoon we went out to the little English artist's cottage. She was so charming and talked and showed us such a bunch of her paintings. She is such a dear.

I bought three beautiful sunny etchings of Algiers. Next time I go I'm going to get a couple of Venetian ones. She colors them so delightfully. You'd enjoy her little cottage and her studio and her tea and cakes and most of all you'd like her pictures. . . . .

I'm so sorry the drive ***[German advance]*** has given you so much anxiety. The folks at home have been so worried. But I don't see any cause for worry on my behalf. You know the hospitals would all be evacuated when danger approached and any way we are a long way back from the front ***[about 65 miles—but with the distinct danger of the Germans driving to the sea between the French and British armies]***.

Yes, I'd be mighty glad to have a real visit with you but I don't see any chance of it till after Fritz is licked. I hope that won't be a great while.

And "our day" will come sometime, too. But I hardly dare to think of that. It seems so far away and yet time passes quickly. No, it doesn't, either. The past year has seemed like ten years at times.

I hope you won't find your new duties too strenuous and that you'll manage to keep well and fairly happy. I'll say good night now but I'll try to add a few lines to this tomorrow.

Sunday Afternoon

We've had Sunday afternoon tea and had two guests. They were two South African sisters that I had met on the train on my way to Paris and found them again yesterday in the course of my travels. Doesn't South Africa sound far away? I've met some charming people from there. . . .

I'm surprised at the little time I find to read.

I hope they didn't leave you long in the Isolation ward of the hospital. [***Erle related later that his first "order" on active duty in the Army—having declined to volunteer that he could operate a typewriter—was administering enemas to dozens of African-American draftees from the Deep South who were suffering from erysipelas, a streptococcal infection. Some Southern Draft Boards, he said, filled their quotas by inducting very ill "second class citizens."***] In my opinion it was a criminal thing to do to place any one totally untrained in such a ward. I see that all the ivory tops aren't on this side of the water.

What is the average intelligence anyway? Occasionally the "so-called human race" seems terribly low to me. . . .

When you reach France it won't take nearly so long to hear from you. That will be a great advantage.

But how I wish it would soon end!!! It's all horrible but would be infinitely more horrible if we don't get peace on our own terms. When the Kaiser imagines he will finish us by reaching Amiens he has a good many more guesses coming.

Did I tell you that I went to Amiens one day to see the wonderful Cathedral there? I did. But that has been some time—perhaps a couple of months since. Bertha sent a clipping about the N.V.H.S. (have I forgotten what those letters mean?) service flag. Perhaps it will interest you so I'll send it on to you. I don't know how much N.V. stuff you hear. . . .

Letter #81

Laura to Erle
April 30 1918

. . . A year ago tonight I left Contagious Hospital preparatory to departure for Europe. Do you remember how excited I was???

The Next Day [***Wednesday, May 1***]

I'm on duty in the Dressing Tent and it is the nearest a vacation I've had since I landed in the land of mist and sunsets and clouds. Just now we are running very light. We dress not more than a couple of hundred cases, give tetanus serum in the afternoon and dispense some medicines three times a day. But it isn't hard work altho it would be tiring if we had many patients. This is the up patient department and there are no sisters—only orderlies—in the wards.

There are two nurses, two orderlies and a light duty patient assigned here so we have lots of help.

I hope your lot falls to some such place. It's much more interesting to the boys than ordinary ward work. . . .

. . . I can imagine the censor with a dull pea green coloring now but he has ceased to be one of my most immediate worries.

My principal one now is the fact that we as a nation can't lay hold of the line and do a heap more than we are doing. It is a bitter disappointment to every one concerned.

But, of course, there was heaps to do more than any one had thot in the way of preparation. All accounts we get of the American fighting men are satisfactory. Thank God for that!

I wish I could say as much of the men in this unit. But some are all right and I shouldn't grouse. It's a very bad habit and one you'll probably acquire. It goes with the army.

I read in the paper yesterday of a Spanish beggar woman (very beautiful) who had prophesied that the war would be over in six months. I've elected to believe her. I always try to believe the things I want to. It's much more satisfactory than the other way about and works out just about as well one way as the other. . . .

Pvt. Baker's memoir:

May 1

Went on night duty in Huts C and D with Miss Anna Olson, a nurse, of Smithport, Pennsylvania. Miss Olson is a very reserved, good-natured, well liked nurse. She always thinks of herself last. "Service" is her motto. Have had an infection on my knuckle for two or three weeks. I keep making it worse by working.

May 2

Slept all day and in the evening walked out to the beach with Pattison. Bought a pair of canvas slippers for 4.5 francs. We were wandering around the sand dunes and ran into the wires surrounding the British [munitions] dump. The wires were stretched just a little above the ground and to the wires were attached tin cans containing pebbles. Of course, when anything struck the wire the pebbles in the cans all rattled. Well when we kicked that wire we were challenged by a couple of guards in just about two seconds. We were able to identify ourselves, naturally, but we sure left there "toot sweet" [tout de suite: very fast].

May 3

Slept all day. Got cold breakfast because cooks have no system for holding out food for night men. Had fourteen hospital ship bed stretcher cases to evacuate before 11:30 and I had to dress half of them. All are glad to go to Blighty.

Our minstrel show made a big hit at the big British 'Y' in two performances.

News from America tells us that the Third Liberty Loan went over big. That shows us over here that our country is behind us and the Allies to a finish. This war is sure getting monotonous. I would like to transfer to some other outfit, artillery, I believe. We look for a great spring offensive now. Both sides are prepared for a deadly fight.

Letter #82

Laura to Erle
This letter's envelope was the first without a British postage stamp in eleven months, the return address marked in Laura's script, "Base Hospital No. 12, U. S. Army" rather than British General Hospital 18. U. S. Army mail was "free." It was marked, as usual on the reverse in Erle's hand, "Rec'd at Camp Dodge May 29 1918".
France
May 7, 1918. [Tuesday]

My dear Young Man: —

And how is the world treating you to-day? . . .

Do you have an afternoon to yourself once a week or is that a custom in your hospital? Do you have a Y.M.C.A.? But you must because you used the Y. M. paper for the note you sent. Do you have entertainments there and do you go to them? How do you spend our evenings? Is your work very hard for you? Do you have to scrub floors as well as wash windows? Does your mail come regularly? Is your food good? How do your officers treat you? Do you have any time at all to yourself? Do you read? or walk? And don't you wish you could put your hand over my mouth and make me stop asking questions? At this long distance I have an unfair advantage but, of course, you may do as you please as to ignoring them. I can think of many more to ask. I'll refrain from going further else the whole letter will be a questionaire. . . .

This afternoon I made new curtains for our room. They are very beautiful ??? and fresh and clean. They are pink roses against a cream and blue striped background. The library table has a new navy blue covering and the window curtain is navy blue and the new matting for the floor has a blue figure. So you see our color scheme has changed.

This afternoon Bess, Emmy and I went walking down to the next village. There isn't much of interest there except an old church. I think I told you once before about visiting it. There is an old Roman Font in it. The mill pond is quite picturesque and would be more so if it weren't so full of slime and scum.

The newspapers don't report much news today in fact haven't for some time. The attempt to get thru at Ypre[*s*] was the last big move. When I read the reports of the treatment by the Germans of the Allied prisoners I get really vicious. I am trying to think up something bad enough to have happen to the Kaiser and his outfit but my imagination isn't equal to the task. But whatever happens to them I hope it comes soon.

We've been getting quite a lot of Americans here this spring but I haven't had a one as a patient. I've not been in the wards where they are. I'm getting a little anxious to have some of them under my care. But that doesn't mean I'm tired of the Tommies and Jocks and Canadians and Anzacs. I like them all.

I am enjoying my work in the dressing tent altho it lacks the coming into close contact with the patients. I miss their talk that one doesn't hear unless one knows them well in the wards. (Gee! but that is an awkward sentence!)

I was pleased to read in the paper the report of the British Labor Commission that has been visiting in America. We have a wonderful country and even if she hasn't been prepared to do much in this offense at least it's a comfort to read of what she is preparing to do. As before stated I want to beat the Kaiser into little bits but I do want to do it quick so we can all get back to normal once more.

I sometimes think that it can't be such a bad thing for us all to get a jolt out of our well beaten tracks. We surely should derive some benefit from the effort individually as well as nationally.

Now please tell me just how you are getting on and don't it at all. I want to know how you are standing your separation from the editorial desk. Don't get all discouraged. Remember that there's a war on now but it won't last forever and that there's a long time to live after it's over. . . .

Do you have entertainments there and do you go to them? How do you spend our evenings? Is your work very hard for you? Do you have to scrub floors as well as wash windows? Does your mail come regularly? Is your food good? How do your officers treat you? Do you have any time at all to yourself? Do you read? or walk? And don't you wish you could put your hand over my mouth and make me stop asking questions?

Letter #83

Laura to Erle
France
May 9, 1918.

Dearest: —

Today and yesterday have been beautiful. Yesterday I worked all day but this afternoon I sallied forth to the second town to the north. I had only been thru it on the train and decided to have a good look around. We thot we'd get a ride but were unlucky going. Coming back we were more fortunate. A Canadian Captain and an Imperial Major stopped their machine and we climbed in the back and sat among a half dozen or so extra tires. You see very few autos are for passengers. They are mostly trucks or ambulances or sort of delivery wagons. The latter type was the kind we were fortunate in capturing.

You see here you walk wherever you want to go and hail any sort of a conveyance that comes along and ask for a lift. It's quite undignified but we have to come off our dignity sometimes.

I wonder if you've started for this side? If your date was supposed to be May 1st you've almost had time to get here. I wish I knew. As before stated—long distance correspondence isn't the most satisfactory communication in the world. . . .

May 10

A crowd came in to play cards and they sort of got me so I couldn't think. Therefore the continuation today. Nothing at all has happened today except that I've been fairly busy and my roommate has been home sick with a cold and I haven't done a thing interesting to tell about.

I've just heard that the lorry has come over from E______[*Etaples*] with the mail and there were many bags. That is for the whole camp but we may draw something from the lot. It won't be sorted at this late hour so we'll have to wait till morning.

Tomorrow I'm going over to see the little English artist again. I want to pay my debts and also get the other etching I have ordered from her. . . .

Sunday Evening

I had to hold this over two evenings before I got the letters. Friday evening's rumor was a fizzle so I tramped away to my little artist lady and had such a pleasant time. I didn't get anything new but just looked and talked and looked and drank tea and ate tiny cress sandwiches and nut cakes. Then we walked back thru the pine woods with some English people—a sister and two officers. I had as a walking companion an extremely polite little English major with gray hair. He had read some topping pronounced tawping stories about girls schools in America and "what are they really like?" Never having seen a "girls school" I couldn't give him much information about them. I did try to explain a high school where both boys and girls were taught the rudiments and also told him of our co-ed universities. He had never heard of either. Evidently the sexes are carefully segregated in England. . . .

Pvt. Baker's memoir:

May 14

Anniversary of our enlistment for service with our government. One year ago today we pledged ourselves to our government for service in the World War. We had a celebration in commemoration as follows:

1. Baseball game in afternoon with Boston Unit. We won 12-4.

2. We had a picnic lunch on the lawn and all enjoyed the company of everyone. Officers, nurses and men all joined in to celebrate.

3. A dance in the evening, 9 to 12, enjoyed by everyone. The dances were one-steps, fox trots and waltzes.

Work is very light. Both sides preparing for action. All patients that could be moved have been sent to England so as to make room for the thousands that will soon come in after the spring drives.

Letter #84

Laura to Erle

"Rec'd at Camp Dodge June 10 1918"
France
May 15, 1918.

Dearest Buddy Boy:—

. . . Yesterday the enlisted men celebrated the anniversary of their enlistment. There was a ball game, a cafeteria supper for the whole unit and a dance at the men's mess. We had one grand and glorious time—especially the men who hadn't been permitted the pleasure (?) or honor (?) of meeting us in a social way since we got down to business. They were surely glad to dance and over zealous to squeeze our hands, etc. and to see us safely to quarters. They're just starved for womenfolks it seems. I was quite surprised to find it so in a camp like this where we all are all the time and fairly free to visit about. (We surely aren't terribly military, I'm afraid.) But last night we all laid aside military castes and everything and just played.

It was one of the pleasantest events of our year—probably the happiest for everyone concerned. . . .

Thursday May 16

We have just returned from a little picnic. We made sandwiches and packed our baskets and went out across the sand dunes to a little pine wood. We took along an alcohol lamp and the coffee pot and my last pound of Uban and had a very gay little time all to ourselves. There were just the seven of us—Emma Matzen, Polly, Frances, Bess Gambee, Conny, Miss Cleveland and I.

The weather has been perfect. It just makes "a fella" want to go out and roll in the grass. Surely the big push will start soon again if this keeps up.

A year ago this minute you and I were sitting in the Tip Top having our last dinner together [***Just prior to the Midnight troop train for New York***^]. Wasn't our taxi ride funny? . . .

You asked if the Hun push has caused me to be worked to death. No, sir, it hasn't so far. In fact just now we have fewer patients in the hospital than at any time since we came. I have had two and three half days every week and several buskshe [*?*] hours off duty. But of course we've cleared out waiting for those Huns to start business again. . . .

THE SECOND YEAR:

SPRING 1918, MAY 17-JUNE 30

"Our cemetery."

VI END IN SIGHT AND POST-ARMISTICE

GERMAN AIR RAIDS CONTINUE

Pvt. Baker's memoir:

May 17, 1918

Big air raid last night. After taps had blown and all was quiet, we were aroused by the air raid signal, the cement factory whistle, and then we heard the boom of a gun fired about ten miles north of us. Instantaneously, out went the lights and everything was plunged into pitch darkness. Soon the buz-z-z-z-z of the great Gotha's carrying their terrible messengers of death was heard. An experienced ear can tell by the groan of the motor when a plane is heavily loaded. Bang!, Bang!, Bang!, went our Archie defenders to the north along the ridge. We could tell by the guns that the whole squadron was coming directly over us.

I am on night duty in two wards where there are about fifty men, many very badly shot up. I went through the wards and lowered every bed to the floor so that the poor fellows would be as safe as possible. The closer a person is to the ground when a bomb explodes, the safer he is, as the bomb blows down and the impact comes up and then spreads out. There is no escape if it lands within a radius of ten feet.

All of the patients who could hobble along at all, I ordered to the trenches which had been dug close to the huts. They took their blankets, gas masks and helmets with them.

The Bang!, Bang! of our Archies during the raid was drowned out by the terrible Boo-oo-mm-m of the great aerial torpedoes which shook the ground just like a small earthquake.

It was not long before the whole sky was full of groaning machines of which there were perhaps twenty-five, and each one was buzzing around spotting targets on which to drop torpedoes.

It required all the nerve I had to act cool as I went through the ward in the dark. The poor fellows were lying there trembling clear through as no one knew where the next bomb would light, and we could hear that awful z-z-z-z of the Hun motors right over our heads. Our brave defenders at the antiaircraft guns had their hands full. Whenever a plane circled round over a battery or a searchlight, the men would dive for their dugouts, expecting a 20 pound bomb or so.

It sure was a great game. As a whole, our patients were well behaved. Only a few showed signs of collapsing under the terrible strain.

During all this bombardment, Miss Olson insisted on sitting on the bed of one young English soldier and holding his hand to keep him quiet. He was scared to death. I just begged her to go out to the trenches but she would not. She thought her duty was to care for the men and she would have died right there sitting by that wounded Tommy if she had to. I certainly marveled, as did all the patients, at her calmness. It was a wonderful example of the bravery of our American girls.

The planes had either gone back and reloaded after having dropped three or four bombs, or else one fleet of planes followed another.

The main object of the attack seemed to be Etaples, the large training area, and machine gun base next to us.

The bombardment covered an area about five miles long and a half mile wide. In this area were crowded from 50,000 to 75,000 troops in training, a few ammunition dumps containing thousands of tons of shells, etc., and ten or twelve hospitals with over 10,000 beds and over half full of wounded soldiers.

Fritz dropped bombs along the main line of the Paris to Boulogne Railway which runs right by our camp, and at Etaples. It seemed as though they would never drop their last bomb and beat it for home. After a plane had flown around a while looking for a good target and dropped a big bomb that jarred the earth for five or six miles, we would say out loud to each other that we hoped to God that was the last one. Everyone remained motionless until each bomb exploded, then relaxed and thanked God it did not land on him. It's a terrible sensation and is about all that I can stand. Part of the time when the planes were right over us, I'd fall flat on the floor, if inside the hut, or try to bury myself deeper in the wonderful open air trenches, if outside, with some of my patients.

Fritz dropped bombs along the main line of the Paris to Boulogne Railway which runs right by our camp, and at Etaples. It seemed as though they would never drop their last bomb and beat it for home. After a plane had flown around a while looking for a good target and dropped a big bomb that jarred the earth for five or six miles, we would say out loud to each other that we hoped to God that was the last one. Everyone remained motionless until each bomb exploded, then relaxed and thanked God it did not land on him. It's a terrible sensation and is about all that I can stand

When the planes moved on a bit, we tried to see them. For a beautiful sight, watch a great air raid at a safe distance. One can hear the humming of the motors up there in the inky blackness, then see the beautiful, weird, powerful searchlights darting all over the sky. Soon the continuous bang of the three inch guns is heard and the bursting shells flash as they explode somewhere near where the gunners think the enemy plane is flying. One can see the silvery streak of the tracer bullets from the machine guns as they try to find the enemy planes and set them on fire. The star shells from the batteries are beautiful when they burst, throwing magnesium or phosphorus into the black sky, thus lighting up the heavens in an attempt to locate the planes.

A sort of cold, helpless feeling passed over me when I saw a plane pass a bright star. It was such a weird sensation.

I was in and out of the wards during most of the raid. Miss Olson sat through it all and we sure admired her courage. She either did not comprehend the real danger or was so thoroughly imbued with her sense of duty, she gave no thought of herself. I think, though, that she too was encouraged by the young Englishman with whom she sat and to whom she talked in low tones. I was sitting near her several times during the bombardment. (Later, Miss Olson paid for her experiences by becoming a nervous wreck. Twice, or three times, she had to leave our area for a complete rest.)

Since I was on night duty, I was fortunate in that I did not lose any sleep on account of air raids as the other boys did. I slept during the day.

May 20

Day began with the big deadly air raid still going on. The people were lying in ditches all over and many bombs were shaking the earth while great flashes flared up from the high explosives. I have never had such a nerve testing time in my life. We stayed in our wards the whole time. They kept the air full of shells which prevented the Hun raiders from coming down low enough to pick out "prospects" on whom to land their packages of dynamite. This raid, like the one two nights ago, lasted nearly three hours and it was not until 1:15 in the morning that the last plane had gone. We were nearly all about "shot" after three hours of mental and physical strain.

The greater portion of the time during the raid I was laying flat on the floor, half over a shell-shocked man who was yelling at the top of his voice. He had been shell-shocked and was completely gone and had a very bad fracture of the leg. If he moved there was danger of a hemorrhage and then things would be all up with him. I had to lie right down on him and hold him still. He kept yelling and was driving the other patients crazy. I talked to him constantly trying to calm him but whenever a giant plane went over and the guns let loose a volley of shells, he would start squirming and would yell, "This one's going to get me", "I know he's going to kill me", etc. Whenever the sickening hum of the German's motor would vibrate through the hut while flying at a low altitude, this poor nerve-shattered, wounded man would scream out and just about pass away. After I had lain on this poor fellow for about an hour, all the time being myself perfectly aware of the dangerous situation, I also was just about "gone".

During the raid, everyone that could buried themselves in the trenches to keep from being splashed with scattering shrapnel. The Hun aviators unloaded their bombs and spent some time in using their machine guns on our gunners and searchlight detachments in an attempt to put them out of commission. At one time three searchlights caught a giant Gotha bomber in their powerful beams. The plane tried to get away by ducking and diving around but the three lights held it until so many shells were bursting around it that it flew down and surrendered. There was nothing else that could have been done for if a shell had hit it, all would have been gone. It landed near the hospital over which it had just dropped some bombs and its occupants proved to be the Commander of the raiders, with a pilot. They were placed in the hospital under guard.

While the raid was going on, a large convoy of wounded troops was coming in and we had no lights on. Many of the wounded had come from the Cambrai sector where there was terrible fighting. I received these men in my ward, lice, blankets, clothes and all. The poor fellows had just come in from being shot and were then being bombed. They were pretty well exhausted from exposure and loss of blood and very easily shell-shocked.

May 21

We all were ordered out this morning at 8:00 o'clock to get busy digging trenches all over the hospital grounds and putting sand bags or earthwork around the wards so as to prevent the sidewise splash of exploding aerial torpedoes, etc. We were given a few hours off from 9:30 to 12:00, and, it being a nice warm day, a gang of us went down to the English Channel swimming. We had a fine time in the water even though the water was very cold.

Had a convoy in. No air raid but we all expected one. There were more than 1,000 casualties from the last one and in all there were around 150 bombs dropped including several torpedoes. Never was a raid more intense. The buzz of the Boches overhead was anything but pleasant, I can say. My nurse, Miss Olson, was very brave and stuck by her ward through it all.

Letter #85

Laura to Erle
May 21, 1918.

Dear Buddy:—

How I'd like to know if you are on your way across or perhaps have even arrived either in England or here! I don't have any mental telepathy nor psychic power in my system so I'll just have to wait till I hear the news from your pen.

This letter, forwarded from Camp Dodge, reached Erle in Chicago June 13 where he was on leave before "shipping out." For a long period thereafter, letters between Laura and Erle took up to months to reach their destinations.

Sat. morning I had a surprise letter from you. It had been just a week since I'd received yours of April 18th and so wasn't expecting one for at least two weeks. I came up to the mess for a cup of tea (Yes really!) and found that big envelope in the H box and then that big envelope rode around in the pocket of my gray gingham uniform for several days.

I'm so glad that you are down (or shall I say up) again to the "adventure" stage. I'm sure things will come along all right for you if you take them that way. You are surely having some experience even if you are in a camp over there. In fact I know you are much busier than we are except when the push is on.

We've been having perfectly gorgeous moonlight nights. It's just too pretty to stay inside the hut. And it's so clear also and warm and summery that it takes me back to other moonlit evenings I've had. These aren't so calm as some other evenings because of occasional fireworks [***German air raids***]. . . .

We are still running very light—just waiting for the big doings to come off. We have extra hours, extra half days and most of us have had a whole day. So don't think of me as being worked to death. I'm not. I'm taking things very easily and lazily.

Yesterday afternoon three English sisters came to us to learn how to give anesthetics. I'll bet they have the shock of their young lives at our mode of living. It will be quite a vacation for them (off duty time I mean) as far as discipline is concerned and I already notice a relaxation on their part. This afternoon they were lying out in the grass in the little squares between our huts and actually had their caps off like the rest of us. Won't they be ruined???

Yesterday after supper we walked upon the hill. The breeze was so cooling. We are really having nice warm weather. It's hot in the sun and that is unusual here. I hope it keeps up a while. . . .

Later

I've been for a little walk and asked one of the English sisters who sat at our supper table and looked lonely to go along. She was quite a good sort and I liked her. On our way back we saw some other English sisters from the next hospital and they stopped to speak with Sister Besley. They asked her where she was stationed and she answered, "I'm over here with the Yanks."

That is what they all call us—"The Yanks." And I sort of like it. There is certainly virility in such a nick name and keeps better than "Sammy" which didn't have much of a run.

English people are after a year among them still interesting and funny in heaps of ways. Of course we are equally so or more so to them. I sort of hope that you're stationed in England for a short time before you're sent across the channel. . . .

Pvt. Baker's memoir:

May 22

Went to bed early after one hour of zig-zag trench digging. Many Boches were over all day taking photos. All went well until 11:00 p.m. when the raid signal blew again. Two or three Boche's flew over and many guns fired on them. They went directly over Etaples and dropped everything between there and Berck-Plage. We crawled into our trenches next to the ward with the "up" patients and waited for developments, but nothing happened. It was all over in fifteen minutes and the rest of the night was spent in peace. These night raids are so uncertain. No one knows when a big bomb is going to be dropped.

May 23

There were about fifteen hundred casualties and four hundred killed in these two big air raids. Most of the killed were buried in one open trench, a hundred yards long. Several nurses and surgeons were killed.

I was so worn out physically, I could not run my two mile race at all. I quit after running a little over a mile although I led the field part of the way.

We finished our trenches. Colonel Collins has a great dugout for the officers and nurses now and he compels the nurses to remain in this dugout during all raids.

Hun planes over every day photographing. Our Archies keep them up four to five miles.

May 24

Fine day. Hate to see night come. Germany is doing all in her power to break the morale of the Allied forces. By blowing up everything possible behind British lines, she intimates she has control of the air and she hopes to force England to quit. England, however, has no such thought, although she is now calling even her seventeen to twenty year old boys into the service.

Lots of wounded men are now coming in gassed in various ways. Tear gas affects the eyes for a day or so. Lethal gas is deadly and will penetrate the gas masks of the soldiers if they are close to an exploding shell. When this gas is inhaled, if the soldier does not die, he will become very sick requiring great care for several months and does not fully recover for years. The gas comes out through the skin all over the body, causing a very painful burning sensation and the skin all peels off. There are three or four other kinds of gases, including pineapple and mustard, which act in about the same manner on the men. Some soldiers come in suffering from wounds caused by burning oil which the Germans had sprayed over our trenches. This is terrible.

Sunday, May 26

The whole British army is being concentrated on their front to stop the German advance around Cambrai and at Kemmel Hill. The Huns started their major offensive to capture Paris yesterday. It seems there is no end to the German manpower. Two great offensives are on at the same time. They are sixty-five miles from Paris now. Big Bertha is dropping her huge shells into Paris about every fifteen minutes. Things look bad for the Allies.

The peasants around our hospital have dugouts over in the sand dunes and the whole town goes over there to spend its nights in safety. No air raid last night but I slept with half my clothes on, with gas mask and helmet both handy, and one ear cocked in order to hear enemy planes.

Letter #86

Laura to Erle

The auxiliary cruiser Moldavia was torpedoed and sunk on Tuesday morning. There were no casualties among the crew but some American soldiers are missing, and it is feared they were killed by the explosion when below deck.

The Moldavia was a twin screw steel steamer, built for the P and O Company by Mssrs. Cairo and Co., of Greenback, in 1903. Her registered tonnage was 9,500 and her speed about 18 1/2 knots. She was before the war a well-known passenger ship in the Far Eastern trade.

France
May 26, 1918

Dearest Buddy Boy:—

Things like the enclosed clipping worry me terribly these days. I wish I knew where you are. If you haven't already started and thereby miss this completely—as soon as you land won't you let me know? Don't wait for a return address or to be settled properly in camp. There has been a little dribbling of mail all week—mostly old stuff and when I remember that your last letter was written April 28th and you were expecting to leave there within a week or so I don't feel very comfortable in my feelings. "Feelings" isn't the word I should use but my vocabulary fails me.

Yesterday a bunch of us walked over the hills to a little village to get dinner. The place was full of our men who had landed but a week ago. It was certainly good to see them and to hear real U.S. voices outside the garden hedge. The last time we were over the Portugase were about everywhere and we didn't like them at all. They stood about in the roads and gibbered and talked. I'm happy to say that we were not disturbed on this trip. Americans are real people anyway.

We couldn't get much for dinner. We had omelette, lettuce salad and potatoes and coffee. It was meatless day and they were out of bread or flour as we took what we could get and glad to get it. The long walk home at twilight was a pleasure. I wish I might have some such with you.

This afternoon we went over to the club for tea and then up over the hill and down again into a delightful little pine woods. All under the trees was clean dry sand with grass and pine needles and a very few wild pansies.

Then we wandered home again and I sat in the mess while the gramophone played dance music and watched the girls dancing and was quite amused and entertained. People are certainly funny things. One dame of about forty years and a face that would stop any ordinary eight day clock—she has been assistant dean of women at a co-ed university for years and came along as an interpreter (superfluous)—has blossomed out, all except the face, into a giddy young heart smasher of eighteen. And she gets away with it pretty well, too. She talks intelligently and dances well.

Another is a child like woman of perhaps thirty five. She is no taller than I and slightly built and she affects short waisted uniforms and infantile ways. But I've known her for ten years and she has always acted so. Perhaps it's natural.

It's a mean thing to do to sit by and watch isn't it? But I do enjoy it. The simple unaffectedness of little Miss Jacques and the charming manner of our chief nurse

The envelope for this letter was marked "Rec'd at Base Hospital No 11 near Nantes, France, July 19, 1918: (evening) forwarded from Chicago and Camp Dodge."

Enclosed clipping marked "Daily Mail"

MOLDAVIA SUNK
TORPEDOED WITH U.S.
TROOPS ON BOARD
SOME OF THEM
MISSING

British Official

ADMIRALTY, Friday.

The Moldavia was a sister ship of the S.S. Monrovia on which Laura had crossed a year earlier.

shine out in comparison. But none are bad. They're simply entertaining whether intentionally or not.

I went to church tonight and heard some very good piano and violin music. I hardly know what it was like because I was thinking of some other things but I did enjoy it anyway as a background for the other things. I don't very often go to church—just once in a long time.

I'm sure I told you about the little picnic we girls had one day last week and I told you, too, about buying the box of 28 prunes for thirteen francs?? Really there's so little to write about these days. I'll wait over another day before mailing this. Maybe I'll get a letter and maybe I'll have something new to write about. Good night.

Monday

Nothing has happened today except vague rumors that the push has started. We won't know definitely till the wounded begin to come in for we always get the news forty eight hours or so late. . . .

By the way I have a new pair of white shoes. I remember how you dislike them so make special note of them. They're canvas and quite flat heeled and look very spiffy with my gray uniform and white apron, cuffs and collar. We all wear white shoes a great deal of the time when the weather is suitable and I believe are quite envied by the English sisters. They are limited to black or tan. . . .

Pvt. Baker's memoir:

May 27

Visited Etaples today to see damage incurred from the last three air raids.

Along the road to Etaples and about two miles or less from Camiers, our base, we came upon what a few days ago was "St. John, No. 7", the fine Canadian General Hospital. It is now a complete ruin. Debris is everywhere. The wounded had been moved to other hospitals. The four hundred dead were all buried in one long grave in the huge cemetery of Etaples for English soldiers, where about 10,000 British already lie. There are huge holes in the ground and wreckage everywhere. The four or five hospitals suffered nearly a thousand casualties. Bombs even fell on a Hun prisoner camp, killing quite a number. Nothing was left of a couple of the hospitals which were full of wounded men at the time of the raid. Those poor fellows were lying in bed recovering from other wounds and then were blown to pieces. It seems a shame. There seems to be no limit to what the Germans will do to come out victorious. If she loses, she will have many terrible things to explain and to pay for.

May 28

Orders came to evacuate all bed patients able to be moved to England. We carried wounded out all day, and tomorrow all "Light Hospital Duty" men are to be sent to England, also. This means that something is to happen soon either the British are going to advance or the Huns are expected to start another drive for the Channel ports.

We have dug trenches all over the camp and around each hut but now we must leave. This area is too dangerous for hospitals. An air raid is a most frightful thing. Getting up at midnight in a hurry and not knowing where the next bomb will light, gives one such a helpless feeling.

May 29

We are all ready for anything next. We have our backs to the wall and the whole British front is lying low to see what Fritz is going to do. Rumors are flying fast as to what America doing and also on every other phase of the tense situation. The British have a lot of warships off the coast here, waiting for an emergency.

Letter #87

Laura to Erle

Erle received this letter, forwarded from Chicago, in Nantes, France, on July 25—two months in transit.

France, Thursday, May 30

Dearest: —

. . . I've had such a pleasant afternoon. We've decided to waste no more time but to explore every available portion of Picardy. We walked and walked and walked, visiting two villages and reaching E______ [***Etaples***] by a new road in time to catch the lorry truck back camp. We met quite a lot of our fellow country men and spoke to them all asking where they hailed from and how do they like France and when did they land, etc. They hadn't been here long enough to be glad to hear a woman speak English. The weather is perfect.

These little villages are pretty as can be and I couldn't help but think what fun it would be to explore them and much of the whole country in a little Ford. We'll do that some time when we get to be millionaires and are living off the interest of our money.

Tomorrow night the officers are giving a dance for some American officers in the neighborhood and I'm intending to go. I may not stay long but I'm going anyway. I don't intend to miss anything any more that offers any excitement or enjoyment. Do you think I'm getting too frivolous?

So much of the time there's absolutely nothing to do and it rains and rains and we can't walk comfortably nor do anything else. . . .

This is Decoration Day and we decorated the graves of the Americans in the cemetery down at E______ . There's about a dozen in all—two officers, two nurses and eight privates. Part have died of disease and several were killed last Sept. in an air raid and some have died of wounds.

I attended a funeral down there yesterday morning—a Canadian nurse injured a week ago Sunday when Fritz paid a visit to this district. The cemetery is beautifully kept but it is so desolate. . . .

Pvt. Baker's memoir

May 31

11:15 p.m. Air raid signal of three whistles and four guns. Immediately we heard the hum of motors and the boom of the guns. The planes came from the southeast and were circling over our district before we got up. The sky was full of searchlights and the shells of hundreds of guns. The flashes of the guns could be plainly seen in every direction and then the sound of the guns long after. The pilots were using machine guns on the searchlights and caused some of them to be put out.

June 1

Air raid was on and in deadly action until 1:15 a.m. Bombs were dropped all around and Fritz came over our heads many times. He dropped many, many heavy bombs all around. He also let loose a very high powered, bright magnesium light which showed the whole district so he could find the way easily. Many bombs were dropped on Etaples with deadly and devastating effect. Heavy barrage now puts real excitement in a raid.

Letter #88

Laura to Erle

This letter, forwarded from Chicago, did not reach Erle at Base Hospital 11, Nantes, France, until August 12, 1918.

France, June 1, 1918

Dear Man:—

. . . We've been having considerable fireworks during the last two weeks but have escaped injury so far. We've little trenches made under our huts and we are supposed to take to them in time of trouble. Fortunately the hospital is running very light just now and we have very few helpless patients. . . ..

Last night at the dance one of the young officers said he thought we wouldn't have to be here long. I asked how long he thot. And he thot perhaps not longer than a year. As he had just arrived he didn't know how long a year sounds nor how long it really is. I might have given him some advice on the subject but I refrained. He was too young and he'll learn. . . .

Monday Evening

Rumors are afloat that we are going to move but not where nor when nor anything definite at all. But everywhere there is a buzzing of conjectures and hints and rehashing of what someone heard someone say. Needless to say we are much wrought up as to the prospect.

If we move you won't know where I am at all. And you might be quite close and never know it. . . .

Pvt. Baker's memoir:

Sunday, June 2

Big raid over Etaples and nearby district. None on us so far. Heavy barrage gives all the appearance and sound of a battle and the sky is full of searchlights and bursting shells—a great Fourth demonstration. We just live in hopes, never knowing what the next hour will bring. I know if a bomb lights near me now I'll be a complete wreck, mentally.

June 3

Fatigue work. I am still in ward caring for heavy surgical cases.

Training for our big field day on the Fourth.

I am very nervous and don't think I will do much running. I don't get much sleep with raids every night, half the night long.

June 5

11:45 p.m. Air raid whistle blew and we noted old Fritz was over Boulogne. We could see all the lights and shells bursting and his very bright magnesium lights as it was pitch dark. He did not come near our camp. The high wind kept him confined to one objective. We were up until about 1:00 a.m. again when the "Clear" whistle was given.

June 6

Raid for two and a half hours again. Fritz dropped bombs all around us but not on the hospitals. A great display. Lost three hours sleep. No sleep again until about 2:30 a.m.

Letter #89

Laura to Erle

This letter, addressed to Erle at Western Newspaper Union, Chicago, was marked "Forwarded from Chicago—Rec'd at Nantes Aug 12—'18".

Sunny France

June 6, 1918.

Dearest Honey Boy: -

. . . You see it's like this: We expect to move. Almost all the patients have been evacuated and no more have come and we are running loose with nothing much to do except pack our trunks and entertain ourselves. . . .We are becoming regular tramps. We start out for any place most within a radius of twenty miles and hail any lorry that comes along and ask for a ride. It's great sport especially when you get turned down. You feel like an eight year old caught doing something he shouldn't be doing.

Today we had quite an interesting time and visited an historic old place and were entertained by a delightful old Scotch artist, Macauley Stevenson. He showed us about and brushed up our history for us (mine certainly needed it) and gave his own and his wife's family tree between times. . . .

I think I told you of my plan for us to tour the country in an auto when we get to be millionaires. This place today is one of the places I want to come back to and show to you.

Were you ever terribly awfully afraid of anything? I have been just naturally scared out of a couple of years growth several times recently. (I'm afraid now I'll never be tall!!) You'd be very ashamed of me and I shouldn't tell you about it for I'd like you to think I'm very brave and wonderful. But I'm not. I'm just a common ordinary coward when anything happens! And the worst of it is that I didn't used to be afraid and now I am. Do you s'pose I'm deteriorating mentally and morally? I'm sure it isn't physical. I don't weep nor yell nor wring my hands but I'm afraid and I can't feel very proud of myself [***an oblique reference to the continuing air raids***].

I'm afraid Francis is letting her affections slip away from McCosh. He's been gone since December and that's too long. There's too many others nearer at hand and then besides he thinks he'll have to live on a farm and she can't see that at all. She's much interested now in an aviator who is only fourteen years her junior. She has met him only three times and one of those times he was loaded.

Sometimes I think it's just as much a tragedy to never grow up as it is to grow old too soon. What do you think? Can you think of a woman of thirty eight chasing out each evening or afternoon and deliberately picking up acquaintances just for excitement? Wouldn't you think it would pall? Heavings!!! And yet she is a perfectly moral person and means no harm in the world. It's just a sixteen-year-old small-

Were you ever terribly awfully afraid of anything? I have been just naturally scared out of a couple of years growth several times recently. (I'm afraid now I'll never be tall!!) You'd be very ashamed of me and I shouldn't tell you about it for I'd like you to think I'm very brave and wonderful. But I'm not. I'm just a common ordinary coward when anything happens! And the worst of it is that I didn't used to be afraid and now I am.

town stunt. It is lowering in dignity but as I said she means nothing by it—just excitement. And Polly is just as bad and she ought to know better. They both need to be spanked.

June 10, 1918.

Reading back over this to catch the drift of my thots when I was writing four days ago I see that I was getting catty and it was time I stopped.

Such a busy time as we are having doing nothing at all! I'll have to go back a bit and bring you up to date.

Friday we went North on lorries and did a little shopping and then came back again, took tea at the club and walked to the sea.

Saturday we gave a tea dansant [?] in the afternoon (I had to work in the morning) and went to a concert in the Y. M. in the evening. The tea was so nice. We gave it in one of the new wards and danced in another new hut next door. It was lots of fun. Frances & Polly invited more English officers over and I helped them entertain them. They couldn't dance much but they tried and seemed to enjoy it. They were particularly pleasant with our Jass band.

The next day being Sunday we gave our regular at home tea. We served it in the new ward we had used as a dance hall and had special music, really high class stuff, furnished by the three of our men. I had a guest—a young Canadian who is in the Imperial Army—and I made Frances help entertain him. He was one of the officers I had helped them entertain the day before and is really a nice, quiet, likable fellow and they do get so lonely and tired of men. He has been in the U.S., has even been to Lafayette, Ind. so feel as if he a neighbor.

That evening we went walking.

Today we slept late, dressed and went out again walking to the second town South of here. Of course we were picked up by a lorry and helped along our way. We purchased shoe strings, tinned milk and a few such necessaries, came home again, made coffee, went out walking again and here I am at eleven P.M. just finishing your letter. The more leisure I have the less I accomplish it seems. I seem to feel that I'm out camping or doing something equally irresponsible. I just want to play all the time. . . .

Letter #90

Laura to Erle
"Forwarded from Chicago—Rec'd at Nantes Aug 12-18"
France
June 16, 1918

My dear Buddy: —

This morning I took my writing pad and walked out on the hill and was going to have a nice visit with you but the skylarks sang and the wind blew and the band played down on the parade grounds and so I didn't get much done except to read over your last letter which arrived Friday night. Then I picked a hand full of daises and came down the hill.

I've sort of been on the entertainment committee this week. Frances and I have been walking and picnicing and teaing and dining with a couple of English officers but now they are gone up the line and we have time to breathe and to catch up our correspondence. They are the first two men that I've met out here that I've really liked and I hate to think of what may happen to them before the war is over. Most of them are such nuts. But we've had a very pleasant week.

Monday [***17th***]

. . . The hospital is filling up again and we are working about every other day. We still hear rumors that we're going to move but no orders have come yet. We've heard that we'll move within three weeks so perhaps we will and perhaps we won't.

The strawberries are coming in this week and they are so nice. They come in baskets. [***scratched out drawing of berry***] (I was going to draw the picture of it but found that my education along artistic lines had been too neglected.) Anyway the baskets are lined with leaves and there are leaves over the top of the berries. The fruit seems so much sweeter than ours. It's frightfully expensive but we buy them anyway.

I'm glad you find the X Ray interesting in a degree and that you are thrown with pleasant people. That is a wonderful help always. Especially does it help in the Army when as many other things irk.

I had thot you'd surely be across the Atlantic by this time. It's really absurd to have the long training there and the way you've been led from week to week for a whole year to think that something is really going to happen is getting to be ridiculous. I hate unkept promises and hot air talks about as much as anything I know. . . .

I worked today but not very hard. I'm getting very soft from so much leisure. Another convoy was in tonight so probably another ward will be opened up soon.

The paper today (with day before yesterday's news) states that an Austrian drive is on against an 80 mile front. I hope they smash themselves all to pieces. I'm so sick of hearing of Germany's and Austria's ambitions. . . .

Sometimes I think it's just as much a tragedy to never grow up as it is to grow old too soon. What do you think? Can you think of a woman of thirty eight chasing out each evening or afternoon and deliberately picking up acquaintances just for excitement? Wouldn't you think it would pall? Heavings!!! And yet she is a perfectly moral person and means no harm in the world. It's just a sixteen-year-old small-town stunt. It is lowering in dignity but as I said she means nothing by it—just excitement.

Letter #91

Laura to Erle
"Forwarded from Chicago—Received at Nantes Aug 12, 1918"—
evidently in a batch as this was one of several marked as received that day.
France
June 20, 1918

. . . We've opened up practically all of the hospital again and I've worked every day this week and haven't had a half day yet. Isn't that enough to make a fellow want to resign from the army? . . .

I suppose you are very busy with your Xray work and with classes, etc. How did the non—com exams come out? As usual?? . . .

Morning [*21st*]

Last night was quite cloudy but today is bright and sunny and breezy. It would be a fine day to walk about five or ten miles with a congenial companion and that I haven't. I think sometimes I'm too particular but I don't enjoy everyone and I'm perfectly conscious of the fact that all people do not enjoy me.

I'm growing a wonderful appetite from this being outside so much. I can scarcely wait from one meal to the next. How are you standing the army fare?

I suppose you noted on the last letter that I am franking my own mail. Of course it goes to Base Censors but I have gotten use[*d*] to that. I do object to the officers in this vicinity reading my correspondence. They all take entirely too much interest in what the sisters write.

The fields about here are red with poppies and yellow with mustard blossoms. Flowers are everywhere. June is such a pleasant month, isn't it? . . .

Pvt. Baker's memoir:

June 24

Air raid again. Beaucoup noise, flashes, etc., for two hours. No bombs on our hospital. We take helmets and gas masks to trenches with us as we have been looking for Fritz to drop some gas bombs just for good luck. The bombs were dropped along the Boulogne railway and on the huge ammunition dump down across the railway, five hundred yards or so from our hospital.

June 25

Another air raid for over two hours. They came in waves and flew around over our heads. Although they dropped bombs all around at different military objectives, they did not drop any on us. These bombs were mostly of the 20# high explosive type, to demoralize traffic and kill troops in camps.

Erle to Harriet Davis, his 21-year-old youngest sister

Addressed to Business Office, Indianapolis Star, Indianapolis, Indiana

Camp Mills

Long Island

Late Tuesday night

June 25, 1918

Dear Little Hettie:—

Your latest letter reached me today from Camp Dodge. It was so fine that I gave it to several of the fellows to read; so fine that I shall take it along to Europe and cherish the message you send me wherever I may be.

We are expecting to leave tomorrow. I regret I have had no time to write at length, but will do so. Address me:

Pvt. John Erle Davis

U. S. Base Hospital No 11

American Expeditionary Forces

France

Letter #92

Laura to Erle

Once again emphasizing the difficulties of correspondence and long waits between letters, this envelope was market "Rec'd at Nantes Sept 16-1918"—nearly three months in transit.

France

June 25, 1918

Dearest Buddy Boy:—

Today I received your letter written May 28th. It came as quite a surprise, all alone—in fact had been traveling about France for some time. I can't understand how in the world the mail service should be so rotten. Surely nothing could be more explicit than the address on the envelopes you send. And yet this one had been to both Base Hospital 21 and British General No. 12. [***Laura was stationed at U.S. Base 12, British General 18.***]

. . . Sometimes I can't see an end to anything, and I get so discouraged. Do you really think we'll ever see the end of this war?? I don't think it agrees with my disposition to have so little to do and to hear a worn out gramophone with some cracked records playing "Gar-rden o-of Ro-o-o-o-osez." The men are surely crazy for music. There is nothing that they crave half so much—except perhaps a Blighty leave—but I mean that they can have. The little Y.M.C.A. Hall is always filled to overflowing whenever there is any sort of entertainment on.

I'm getting to be about as bad as the men. I want to be entertained and I've been going to all the concerts and movies for two or three weeks—perhaps four in all—awful dissipation, isn't it?

I imagine from what you say that politics are the same in your unit as in ours. I've long since given them up as absolutely hopeless. It was fine that your friend Ford put something over on them. I wonder if other sections of the army are run on the same principles as are the medical units? But at least the officers have special training for their commission in the O.T.C.

By the way, did I ever tell you that my pet orderly, Art Hall, has at last been made a sergeant and is beginning to hope for the opportunity to transfer to Infantry or some other fighting force? He is a good man wherever they put him and he should have the opportunity to transfer. . . .

Just now I'm peeved at Miss C. I had a note from a

Scotch Canadian officer [***presumably Lt. Murray***] asking me to go to tea with him or if I couldn't do that to "take him walking" this evening. I passed it over to Miss C. and asked her what she thot of it. She said, "Oh it depends on whether you are crazy to go out with a man." Catty, don't you think? The Lord knows I've never shown any great interest in any man except one with whom you are well acquainted and I felt as if I'd been slapped. Just for that I'll go walking and any time I ask her advice I'll have changed considerably. But maybe she was just tired. And the man is just lonesome. He is just about as much "interested" as I am from a personal point of view.

We've had no hostile visitors [***German bombers***] this moonlight week as we did last month. We can get along nicely without them. . . .

Letter #93

Laura to Erle

"Rec'd at Nantes Sept 16, 1918". Letter on Australian Red Cross stationery note paper, undated but calculated from July 2 postmark as begun June 30, 1918.

Dearest Buddy Boy:—

Did you feel me close beside you yesterday afternoon? I had been playing tennis—yes, in the war zone—and dropped on my bed for a few minutes rest before going to the ward. I had the weirdest sort of dream—of you and woke up smiling. I'm still smiling. Are you?

We are having some more glorious weather after a very few cloudy days and I'm trying to get just as much out of it as possible. I've walked by sunlight and moonlight, played tennis, worked a little, danced a little, have gone out to dinner a few times, lunched on strawberries and coffee furnished by my neighbors and have done about everything except write letters.

We aren't at all busy but all the wards are open and waiting. Our ward is only half full and the cases are very light. There are three nurses and we have lots of extra time off. . . .

A hospital of this sort is infinitely more cheerful, more hopeful and cleaner than an ordinary civil one. But at that you know hospitals aren't so sad as the people outside imagine.

But I think most of the men have the same sort of religion that they had before the war. They don't go to church often but they are far from being Atheists. Perhaps their non-church going is caused by compulsory attendance of services in the army. And when they get away from the battalion they want to rest up from it.

What do you think of the Austrian failure? I wish it were Germany also.

Monday [July 1, 1918]

I didn't finish this yesterday as I had intended. I was too busy. I worked till nine-thirty and as the work was all finished then and the two other nurses had returned from mass I left the ward and went to the village for a basket of berries for the men. Then Miss Horowitz (I'll tell you about her sometime) and I went down to the village

I think most of the men have the same sort of religion that they had before the war. They don't go to church often but they are far from being Atheists. Perhaps their non-church going is caused by compulsory attendance of services in the army. And when they get away from the battalion they want to rest up from it.

church to the services. The music was atrocious but I heard very little of it anyway I was so busy thinking. After church we walked up over the hill and back to camp just in time for dinner with strawberry shortcake—um-m! After dinner I cleaned my white shoes and went out to the ball game. It was between the hospital unit team and some other Americans stationed near here. And another American camp sent its band over and we were entertained by real jazz music. "Katrinka" and a few others of like nature took me back to the Rienzi.

Of course I sunburned my nose most horribly but that is to be expected.

Then I went walking down past the little lake with a Scotch Canadian officer who has spent most of his time before the war in the woods of British Columbia as a timber contractor. He's a queer chap but "lonesome" as most of them are. Then I came home to supper and went to church. After church Emmy and I walked up a little ravine between the hills in the twilight. It was so peaceful and quiet but before we were asleep we got up again to watch the fireworks [***night bomber attacks***]. I'm glad to report that I'm not so frightened this time as I was last month. After all it's just a state of mind, isn't it?

This morning I slept late and came on duty about nine-thirty, did the few little dressings there are to do and then sat down to write to you and to listen to the victrola between times.

Gee! I wish I could sketch. I'd love to just to show you this group of Tommies about the victrola. . . .

Today our American canteen opened and you should have seen the queue (sp?) waiting for cigarettes, matches, candy, pickles and coffee. I bought 4 lbs of coffee and so did my roommate so can imagine me as a real coffee fiend when you see me next time—trembling hands, nervous wreck, etc. . . .

Evening

I've just come from a Y.M. entertainment—elocution by an ex-padre—absolutely the worst & funniest I've ever seen. Thank goodness I didn't take up elocution as a life work!!!

Northwestern University student jazz band that entertained all around the area.

Summer 1918, July-September:

AMERICANS IN FORCE AND THE TIDE TURNS

Pvt. Baker's memoir:

July 1

Raid. Loss of three more hours sleep again. No bombs on us but near. Ground shook with their resounding vibrations and concussions. Oh, for a gun to protect one's self, even though it wouldn't, it would feel better on our nerves. Mine have been tried many times. I sure do admire the nerve of our nurses and wounded men. No one knows who is to be next. We say if one of these bombs has our names on it, we will get it. That is the British fatalism idea; however, it gives one courage to "carry on" till hit under fire.

July 3

Feeling rotten. Loss of pep which I need for our long grind—the two mile run. I can't rest at all. I sleep two hours in the morning but I cannot make up the loss. I keep going because I cannot sleep soundly and I cannot sit still. Gee, what I wouldn't give for a week of good sleep away from these damned Huns.

July 4

Track meet with British Army at Camiers, our own camp.

The great Fourth. All set for the big day. I am in a physically weak condition just when I need all the energy I can get. I have no ambition. Loss of so many good nights rest put me on the bum. I never saw a more mixed crowd of Allied people than at the meet. All went fine but I had to drop out in the mile at the 3/4 mark and at the 5/8 mark in the two mile event for lack of stamina. I never felt so weak in running before. Had a fine dinner for all afterwards.

July 5

Still feel very weak but am soaking up all the sleep I can get during the day, so I can miss some nights.

The 30th Division of Yanks and Australians are pushing back the Huns and captured Le Hamel at the Amiens Apex, which is only twelve miles east of Amiens. Our boys did wonderful work and the Australians and Canadians are very happy about it. We feel more secure now and our morale went up about 500% through this wonderful drive. All of the Allied troops have picked up their lower lips and are eager to march on to Berlin now with confidence.

Last night I was out walking across the dunes with a rather oldish Scotch Canadian who has spent the greater part of his life since he has been grown in the woods of British Columbia as a lumber contractor. He isn't very used to women and I don't believe likes them very well—thinks men are much better and squarer, etc.—but he doesn't make friends easily and is very tired of everything concerning the war... He has just lost a young brother over here and is quite broken up over it. But you know it is really men like him who are doing what is done here. He is a real man. . . .

Letter # 94

Laura to Erle
France
July 7, 1918.

Dear Buddy:—

. . . In the evening the M.O.s gave a dance and we had a very pleasant time altho a crowded and much stepped on feeting seemed to be universal. My white shoes were almost ruined.

At the Sports I met the nicest boy. He is from Des Moines originally but being in Scotland at the time of the outbreak of war has had nearly four years service in the Imperial Army. He is a flier (Royal Flying Corps) and has red hair and a freckled nose and a very pleasant manner. Frances quite lost her heart—for the hundredth time since we landed here. And then he is only about fifteen years younger than she is but what's that in a lifetime?

He is the second man out here of whom I have really approved. The other was a Canadian in the Imperial Army. I've met heaps of men but I find I'm not very keen about most of them. But I'm beginning to think that you didn't properly appreciate me in Chicago. Huh! I had no idea (and I'm sure that you had not) of the great privilege it is to walk with, talk with and mostly to a fairly intelligent (??) woman who speaks and understands English. The officers who have been up the line a year or so or the whole time of the war are so weary of the sight of men and khaki that anything else—even a sunburned, red haired person—looks good to them.

I'm getting really frivolous! I scarcely ever miss a concert or a movie (known here as cinema) even when they occur as often as once a week. And then I go to tea at the Club and then I walk about on the sand or over the hill. Such desperation!!! Ain't it awful?

Work is still very slack and we are expecting to move sometime this summer either within a month or as soon as possible. It's all too indefinite to get excited over. But if we do move you won't know where I am and you might be located quite near and never know it. . . .

Honey, I wish you could have seen the sky a few nights ago. The half moon was just sitting on top of the hill and the search lights all over the sky. It was well worth seeing. I've never seen just that particular view of it altho many variations.

There's a Cockney just waking from ether and he is very loud about it. He is very profane and sounds just like a pirate out of a ten center [movie]. Englishmen swear very differently from Americans. "Bloody" is an awful oath and "Bugger" or "Booger"—don't know how to spell it is a terrible name to call any one. To me they sound very mild but the Tommies are trying their best to quiet him so as not to offend my ears. It is in reality entertaining and I'm having quite a time to preserve my dignity and not laugh. . . .

Last night I was out walking across the dunes with a rather oldish Scotch Canadian who has spent the greater part of his life since he has been grown in the woods of British Columbia as a lumber contractor. He isn't very used to women and I don't believe likes them very well—thinks men are much better and squarer, etc.—but he doesn't make friends easily and is very tired of everything concerning the war. He has simply a great longing for quiet and his woods and trout streams. He has just lost a young brother over here and is quite broken up over it. But you know it is really men like him who are doing what is done here. He is a real man. . . .

Tuesday morning [*July 8*]

My dear, I didn't mean to be so pokey about mailing this billet deaux (sp?) [***billet doux***: ***love letter***] but I laid it aside till I could find Edna A's letter. I can't locate it anyplace so I'll send this along. Perhaps you knew about the Dills anyway.

Last evening Frances and I had planned to go over the hill to a village for dinner with the Scotch Canadian before mentioned. The weather seemed to be so sickeningly warm and close which is very unusual here that I refused to go so instead we went down to the little Hotel du Lac in our own village. While we were waiting for the meal to be prepared we went over into the pine grove beside the lake and sat on the sand and cooled off. We had a fairly pleasant time of it even tho we were rained on and quite soaked on the way home.

This morning I slept late and came on duty about ten thirty—such industry!!

And now I'm having the afternoon off. I'll take my book and go up on the hill and meditate on my sins and shortcomings. Do you think that an advisable way to do or should I throw aside all responsibility for past misdeeds and dream about the future apre le guerre? *[**après la guerre**: after the war]* **. . .**

CHICAGO BASE HOSPITAL 11 TO NANTES, LOIRE, FRANCE

Letter # 95

Laura to Erle
July 11, 1918.

Dear Man: —

I haven't a thing in the world to talk about. Life is moving along very calmly, the days pass, we aren't busy at all, we eat, we walk, we talk and I think that is all. Also lately nothing new has happened and yet I want to talk to you so much that I'm going to make an attempt to write.

Your last letter mailed June 6th reached me about a week ago. Thirty five days since I heard. That's only five weeks but it does seem so long. And, m'dear, it's very nearly fourteen months since I saw you. I do hope it won't be that much longer yet before we meet. . . .

Misses Gambee, Pawlisch and Matzen have gone to Paris on leave. Miss Cleveland has gone out walking with a bunch of girls. Miss Hampton has gone out strolling with an English officer. . . .

There's a tiny moon tonight and a few clouds and several searchlights. Once we heard the raid signal but nothing happened. I suppose some one had the wind up because the "all clear" sounded in about five minutes.

This is the first letter addressed to Erle in France: Pvt. John Erle Davis/Base Hospital No. 11/ American E. F./France, address overwritten "Nantes/Loire/ gnfre", "rec'd July 29, 1918 at BH No 11 France". It took only four days to reach its destination. Note that because this letter arrived out of sequence—earlier letters were not delivered until August and September— there was no 'proper" prior introduction of Laura's "Scotch-Canadian" officer friend.

Saturday [*July 13*]

Last night came your letter of June 17th and the note mailed June 26th on Long Island. Sixteen days should allow time for the voyage so you either are already landed or are nearing land. It is hard to believe after the long, long wait. And now I await the announcement that you have arrived. I do hope we'll be able to meet. I'm due a leave and maybe I can manage to see you if you are near Paris or anyplace where I could get transportation. . . .

Miss Urch has at last been transferred to an A.E.F. mobile unit and we are all so glad for her. And the transfer came thru right over the heads of the powers that be in the unit!!

Last night we girls had a party at the mess for her. We just sat about and played cards or talked and three of the boys came up and played the piano and violin and sang and we had such a pleasant time. It was the nicest little party we had had since we've been here—and no men except the musicians were present. The latter enjoyed it just as much as we did I think. As no officers were present they were just part of the crowd and that pleased them, too. I don't think they had to think once that they were privates. When it was time for speeches one of them got up and made the nicest little talk saying how much they would miss Miss Urch and how they had all felt her influence more than she could know. Honestly I could have hugged that boy! He did it so nicely and sincerely. . . .

I'm sitting out on the hill writing this and every little while the wind whisks the paper away from me. I suppose it is a hint that the day is too pleasant to spend writing. Anyway I must run down the hill and get dressed because I'm taking that Scotch Canadian to tea at our club and he is taking me to dinner this evening. . . .

Letter # 96

Laura to Erle

The envelope for this letter was signed "Robert Murray" [Laura's Scotch-Canadian officer friend] on the front. Such an officer's signature bypassed the normal unit censor. Addressed to Pte. John Erle Davis, it was marked "Rec'd July 30 1918 at B.H. 11—France"

France

July 16, 1918

Dear Buddy Boy:—

. . . Tonight or rather this afternoon Conny and I are walking over the hill to the next village for dinner with a couple of British officers. And to avenge myself on the one of them—not being the right man I'll make him frank this letter to you. Whatever happened to the ones I sent to you not in green envelopes? Please don't forget to tell me. The last one I had this officer frank for me so it should go thru without being opened. I hate to have strangers to say nothing of the officers I know reading my stuff. But I do recognize the necessity for a censorship. . . .

Day before yesterday at our tea I sat at the table next to a Canadian officer who had been to North Vernon. Just think of that. I almost arose and clasped him in my arms but restrained myself lest he be entered on the lists as a war casualty. Honestly I was pleased.

I've never met anyone before who has even heard of it. I've seen only one Hoosier since I came except for the few in our own unit. . . .

Pvt. Baker's memoir:

July 17

News from Rheims front shows the Huns have gained little. There are now eleven U. S. Divisions along the front.

July 18

Hun drive reported stopped and big counterattacks are in progress. American 2nd Division with Marine Corps met Huns at Apex, twenty miles north of Paris. Big battles at Belleau Woods and Chateau Thierry. Rumors are wild.

Letter # 97

Laura to Erle
"Rec'd July 30 1918 at BH No 11 France"
France
July 18, 1918.

Dearest Honey Bunch

. . . It is so comfortable to know that at least you are on this side of the Atlantic. You'd have been so cheated and I'd have felt cheated, too. I do so want you to know something of my own experiences even tho they amount to very little in the light of what some others are doing. You see, I just can't quite get over the idea that I'd rather be a man. But then if I were you wouldn't love me. Maybe you won't anyway for very long and then I'm going to drown my troubles in vin blanc and cigarette smoke!!!

How do you like the English, or rather European custom of women smoking? It surely looked strange to me at first but now I don't mind.

Tomorrow I have a day off and I'm going up to B______[***Boulogne***] shopping. I'll look at everyone most carefully lest I pass you by unnoted. There is a possibility that you won't tarry long on English soil. We were just nine days in England, eight of which I spent in London. And when Conny came they were just one day in Liverpool.

The heading of your letter paper was clipped off [***by a censor?***] so there wasn't even a date but I supposed from what you said that you landed in Liverpool and meant "hills of England" but as even the post mark was blurred I don't know a thing. And of course all written on the back of the heading part was clipped too (Strange that two sides of the sheet should have been clipped!!!). But, sweetheart, just imagine me getting a letter from you within two or three days of the time it was written!! Why that's enough to make me delirious! . . .

I'm not going to try to finish this tonight because I haven't a great deal to say and also because I don't want to ask that Scotchman to frank another one so soon. Perhaps a green envelope will appear in the offing and good night.

Sat. the 20th

I received your post card saying you were still moving. As it had no date nor heading (such as France or England) I'm very wise. Surely your most careful censor, Capt. Richardson, will allow you to mention the name of the country. If not please put 'Eastern Hemisphere' so I'll know you haven't gone back to the States. And a date shouldn't be blotted out.

I didn't see any stray American yesterday in B______; in fact most of them are with the Am. E. F. now and I'm glad of it. They are so much happier, better fed, etc. and then Americans are different from other people.

Are you accustomed to hearing yourself called "Yank"? It was funny at first but now I never notice.

Sunday

Today, I went and signed up for a leave. It's worse than useless to think about seeing you unless you'd be near Paris or else in our vicinity and I know I'd not know where you are and everything is all wrong today. . . .

We think now we'll go to Chamonix near Mt. Blanc for a week in the French Alps and perhaps a day out at Lake Geneva. Won't that be grand and glorious? And then the next vacation I expect to spend partly with you either at Nice or Paris!!! Such boldness!!! Even an hour would help, wouldn't it? . . .

Now please, honey, if you are settled yet let me know if there is any chance of meeting me in Paris. I know the chance is slight and yet there is a chance. We won't leave here till August 1st and will be due back on Aug. 11th. Of course we go thru Paris going & coming. . . .

I'm going to ask the colonel to find out where #11 is to be located . . . .

Nantes Cathedral, France.

Erle to his mother

[*"American Red Cross/Croix Rouge Americaine" stationery*]

France, July 24, 1918

My dear Ones —

Have you been worrying about my safety? I trust not. As soon as our ship reached this side of the Atlantic I mailed a little letter home. Before leaving America I filled out a post card which should have been mailed to you from New York as soon as it was known that our command had reached Europe. Did you get these notifications?

We have been in our own hospital camp about a week now. It is situated just outside the city of Nantes, on the grounds of an old estate; and the neighborhood is naturally beautiful and quite picturesque because of the buildings and people. Our buildings are portable barracks of wood; made in America, sent to France in sections and bolted together on the grounds. After things are arranged properly the place will be much more pleasant for us than Camp Dodge was. So far, we have worked hard early and late cleaning the whole works—cutting brush, raking, scrubbing, digging ditches, making roads, and so on. Evidently the labor hurts nobody, for there's not a sick man in our entire command. Myself, I am highly colored and abundantly freckled.

Coming from the States, we traveled over an unusual route — a surprise to all of us and very interesting. We passed through half a dozen large cities and made several stops at rest camps for a day or so each. When we finally arrived here many of us were sharply disappointed to learn we were not closer to the battle front, where things are happening. But this disappointment has been dissolved to a fair degree as we have become better acquainted with the surrounding territory.

Sunday I spend in Nantes, going to high mass at the church of St. Nicholas and to vespers at the Cathedral. St. Nick's is very large, of modern gothic; a handsome pile to look upon from the street, and inside really magnificent in structure, architectural balance, coloring and decoration. The singing was Gregorian (which I don't care for), but the incidental organ music was very good. The Cathedral looks quite moth eaten outside—it is a 15th century structure, with two square stone towers—but inside it is in fine condition: all ivory white touched with soft colors lent by the stained windows. I'm enclosing pictures, but they are very flat indeed by comparison with the real thing. . . .

By the way I had dinner Sunday with a French lieutenant of infantry, returning to the front after a short furlough. I was in a hotel trying to change American money for French, and was having a hard time making them understand what I wanted when he stepped up and offered his services. Last evening I went over to a little restaurant in the village near by for supper. It lasted two hours; I met a couple of French soldiers there, and they and the old Madame who runs the joint made the occasion a sort of party. I am invited to have dinner with them as their guest in the family's private garden Sunday, and trust that I'll not be too busy to go.

We have thrilling stories told us by some of our patients just back from the fighting front. I had occasion to interview several of them this afternoon, and it was great stuff they told me. It seems to be the opinion up on the Line that the Germans can't last much longer if the Yankees and French keep up their drive as it has been going the last two weeks.

I haven't any more paper now — besides our officers (who censor our mail) have asked us not to write very long letters but to make 'em short and write oftener. . . .

Letter #98

Laura to Erle
France
July 27 1918.

Dear Buddy Boy:—

Why haven't you written me this week? Are you still on the move and are you moving to Italy, to France or even to that far away Russian place Murmur [***Murmansk***] or some such name? . . .

Next Thursday I go on leave to the French Alps. I hope I get that letter before I leave else I'll go feeling very much cast down in spirits.

Last night we had a dance. I invited the Scotch Canadian—mentioned several times before—and he surely had a good time. He acted like a kid with a new toy. I suppose after I've been in the front line for four years—I'm back a long way from the front line—such a thing as a two-by-twice dance, but with excellent honest-to-goodness Jazz music, will make me think I'm having a wonderful time. I have a pretty good time most of the time anyway.

I've gone to more teas and dinners and walks, etc. this summer than ever in my life before—but we aren't busy and must amuse ourselves some way.

But our hospital is filling up pretty rapidly now and we haven't so much spare time. But our cases are still light and that means less work than the badly wounded make. If we don't get someplace where we have real cases I'll be so lazy that you'll think I'm past redemption.

I haven't done a full day's work—that is to really work all the hours on duty—since March or perhaps April. I did have a few strenuous nights in April. . . .

. . . I've been so proud of the reports of the American fighting qualities that I can hardly keep from swelling up and expanding my chest as I walk about. But it will hardly do to be quite so exuberant among the conservative Britons.

The writeups in the English papers are very complimentary.

Did you notice a little note in one of them—perhaps the Express—concerning the dislike of Yanks for the English coffee and salads? I agreed but I'd have put in a longer list of English non-edibles.

I was really hungry for something really good all the time I was in London. Did you have any trouble satisfying your appetite? But I suppose you were rationed on American food all the time? [***Erle's A.E.F. unit crossed the Atlantic in an English ship, and they "enjoyed" English rations.***] We lived in a hotel.

The French can certainly cook. I really enjoy their food.
. . .

Letter # 99

Laura to Erle
Franked by Robert Murray on the front, the envelope for this letter was marked on the rear, "Rec'd Aug 8, 1918 at Nantes"
France
Aug.1, 1918.

Dear Buddy Boy: —

Two days ago I received your note from Nantes saying that you were located there and were very busy

I quite envy you your location. I do so want to be in a hospital on an old estate, or in an old convent or monastery or even a hotel. Everything here is barren and sandy and camp like. We've heard nothing recently of any movement orders so perhaps we'll stay right here for duration.

getting settled. It lifted a big load from my mind as I had about decided that you hadn't reached France after all.

For some reason the note written on July 19 & mailed on July 22nd didn't reach me till July 30th. I wonder why the mails are so slow. But it is so wonderful to hear from you within a week or ten days period that I shouldn't object.

Are you taking patients yet? And if so what are you getting by way of cases?

Ours are still very light—but of course there isn't a great deal doing on our part of the line.

We expected to go on leave this week but we didn't get off. It will be two weeks now before we get away.

I imagine you are having exceedingly warm weather in Nantes. It is uncomfortably warm even here this week. Last week we had one shower after another but now the sun beats down from morning till night and even the breeze has almost abandoned us in the daytime.

After sundown it cools off rapidly and a wrap is needed outside. . . .

The papers report Allied progress daily but I realize that the war is far from finished yet—unless, of course, some unforeseen event finishes it.

I quite envy you your location. I do so want to be in a hospital on an old estate, or in an old convent or monastery or even a hotel. Everything here is barren and sandy and camp like. We've heard nothing recently of any movement orders so perhaps we'll stay right here for duration. [***Erle's Base Hospital 11 was located on the grounds of the Chateau du Grand-Blottereau, an estate on the outskirts of Nantes, a city near the mouth of the Loire River, southwest of the Britanny peninsula, several hundred miles south of Laura's location.***]

Are you too far south for air raids? I rather think so but can't be sure.

We are not but I'm happy to tell you that I have my nerve back again. I was so ashamed of myself this spring. I feel pretty sure that it is a matter of faith. I sort of lost faith this spring and I'm sure that it effected my nerve and courage—if I possess any of the latter quality.

I thot just then I heard a "Jerry" motor and went out to look but could see nothing. I had just come back to the table when the anti aircraft guns began so my ears were right after all. There is a different hum to the German motors. It does sound so vicious at night. It reminds me in a way of the horror I had when a little girl of the buzzing of a wasp or bumble bee about my head. I was desperately afraid of such things at one time. But I think we are far enough from good targets to be damaged a great deal so please don't worry. I am all right. . . .

I'm not planning on having this letter censored as I have a very good friend who will frank it for me without reading it if I'll just promise to tell nothing of military importance. . . .

Pvt. Baker's memoir:

August 7

We were very hungry and tired when we got back [from leave]. I was ready for a good sleep. Am assigned to Capt. Krost's ward again, which is full of very heavily wounded men. He, Jardine and I get along fine. I'd rather be working for heavily wounded than at anything else. I feel I am at least doing some good.

August 10

About sixty of the girls from the Women's Army Auxiliary Corps of the British Service came over to our hospital and we had a dance and a feed in our mess hall. All over at 10:00. Big time.

10:30 p.m. Big air raid for two hours over Boulogne and Etaples. Saw three different night riders get caught in the searchlights but the gunners failed to hit any of them. Sid Barteau and I both hit the same hole at the same time. It was not big enough for both of us but we sure got close to Mother Earth when a shell came down close to our location—a dud. Wow! What a shock. We now have tin hats and always wear them.

Second Leave: to the Alps

Letter # 100

Laura to Erle
Chamonix
Aug. 10, 1918.

Dearest: —

I shan't try to tell you of all the beautiful things I've been seeing this week. Suffice it to say that we left Camiers at 10 o'clock Tuesday morning and arrived here at 10 P.M. Wednesday [18 hours traveling]. We made good connections except the latter part. We were held up for six hours at Lake Annecy but that wasn't a hardship at all as we improved the time by a good lunch and a boat ride around the lake.

We are high in the French Alps very near both the Swiss and Italian border. In fact we took the car up to the Swiss border and walked across the bridge and mailed some cards from the other side. That is forbidden but we seemed to be so much of a curiosity to the guards that they let us do it. Perhaps, too, our faces were guileless.

I've tried so hard to get some decent pictures here but have given up. The photos or prints are so inadequate that they are only a disappointment. But we did find a little shop where a French photographer had some beautiful photos of Northern Africa—Egypt, Algeria & Morocco. He has some wonderful poses and distances & lights and shadows in his photos. I know you'll like them immensely when I show them to you (Wonder when that will be.) I seem to always fall for those African scenes. Perhaps formerly I was an Arab or Egyptian Princess. I'm sure no one ever thot that in a previous existence they were a slave or a beggar??

Today we went up on mules to the tip top of Mt. Brevent and looked across at Mt. Blanc. We walked most of the way down and it was something over six miles the guide told us. Personally I'm sure it was ten.

Everyone is so pleasant to us and anxious that we should have a nice time. In fact it is as charming a place as I've ever visited.

I received your letters of July 25 & 26 just a day or so before I left camp. When I read it I was quite jealous of you to think of the way you meet real people. No doubt you've had more experiences in your month here in France than I've had in all the time I've been here. I do wish I had been born a man. That is I wish it very often. Of course I don't always want to be a man because perhaps you'd love some one else then. And you know I wouldn't like that at all. But men do have so much pleasanter a time.

No, I don't speak French at all. We have so little use for it situated as we are and I very much disliked the only available teacher so refused to continue to study. Conny does wonderfully well and she manages to get us about the country on her small vocabulary.

You know our part of the country has been in British hands so long that most everyone speaks a little English and we have no use for French at all till we go on leave. . . .

I'm going to bed now because the mule ride has made me very sleepy but I'll try to add a note and send it along tomorrow. We expect to leave here about Tuesday for Lake Geneva—just a day—and spend another day in Paris. Then back to camp. Goodnight, sweetheart. How I'd love to put my arms around your neck and kiss you goodnight! (Bold girl!)

Sunday Evening [***Aug. 11***]

My deah, did you ever walk down a mountain for six miles and then walk about the next day? It surely does take an effort to walk. I go up and down steps just like a rheumatic dame of eighty odd summers.

It is so beautiful here. The town, which is really just a little summer resort of family hotels, is down in a valley and all about are the Alps—some snow covered and some all covered with the trim, straight evergreens—cedars or larch. Just at the edge of town is the prettiest grove (known here as the forest) all carpeted with moss and the brown needles from the trees. A little rushy stream flows thru it and down the valley. And people are so friendly and folksy. There are many English speaking ones even among the French. They are the first civilians I've talked to of course with the exception of the little English artist. Every one I've met before has been military.

There is an American woman and her fifteen year old daughter whom we like very much. Then there is an elderly American widow who is on leave from nursing duties in a French Hospital at Cannes (Deliver me from French Hospitals if this is a sample of their nurses) and a nice little English woman who is married to a Frenchman. They are our best friends here but everyone is anxious to talk to us and be nice and give us information. It has been a rest and a change. Now if I could come back past your camp and say, "Hello" I'd feel as if I'd really had a leave. . . .

Letter # 101

Laura to Erle
Back in Camp
Aug. 17, 1918.

Dearest Buddy: —

We returned yesterday P.M. and found everything just as we left it. The hospital is fairly busy but no heavy cases. I was given morning hours so I ran away from my roommate and came over to the club to answer a few of the letters I found waiting for me.

I found ten letters, a post card and about a dozen magazines stacked up on my bed and maybe you can imagine my delight. I didn't even wash the inch deep grime from my countenance. I waded right thru those letters. One important piece of news contained in them was the announcement of the arrival of a brand new baby girl in my brother Silas' home. It is the fifth but of course just a perfect baby as all of them are. Silas is barely able to get about on crutches but of course in a year's time he may be back to normal. His arm has healed nicely but his knee does not function properly. Nellie [***wife of Silas***] has surely had a terribly hard year.

The folks are getting on nicely. Pa hasn't been working for a year and has just put in his time in the garden, running the Ford, going fishing, etc. I know except for Silas' condition they are having a better time than ever before. This spring Pa had an increase in his [***Civil War***] pension and with what little else he and mother have they can take things easily. It is time surely. He is 70 and mother 66. . . .

You want to hear about the Scotch Canadian. A biography? He left Scotland for Canada when about nineteen. He farmed, tramped, fished, gambled, prospected, logged—what else do western Canadians do?—for twelve years, living much alone in the logging camps, etc. He enlisted in the Canadian Army sixteen days after war was declared, came out here in 1915, was promoted from private to a corporal, corporal to a sergeant, won a military medal, was recommended on the field for a commission, took out his commission last September as a 2nd Lt., was shortly made a full Lt. and is now acting Captain but refuses to put up the third pip on or to call himself captain till he is a real one.

He is not very large, has dark hair, hazel eyes and a ruddy Scotch complexion. He uses more or less American slang but his rolling r's and Scotch vowels give away his nationality. He hasn't had a great deal to do with women, is rather shy, is quite well informed, has read much, (largely self educated I think), loves music and the open, is a little rough but a real man and I can't think of another thing about him. Is that sufficient to satisfy your curiosity?

We were a day and a half in Paris and met a girl on the street whom we knew. She was from Unit 14 [another Chicago base hospital] and was in town just the one day on her way to their hospital. She was a special friend of Miss Cleveland.

It was too warm there to really enjoy ourselves. We went to the Opera Comique to hear Carmen. It was quite enjoyable altho there are other operas that I like better. We stopped there to do a little shopping but the shop was closed because of its being a fete day. So we came home somewhat richer than we had expected to. But I still need some articles of wearing apparel! I'll have to hop a lorry and go to B______ and see what I can do.

It's nearly noon so I'll stop now and write a note home. I'll finish this tomorrow.

I have a green envelope anyway and there is no one in the offing to frank it for me. Scottie has probably gone to S. [***Scotland?***] on leave or else up the line and I know no one else well enough to ask the favor. It is a favor, you know.

Evening

I've just come in from a concert at the Y.M. It was all music and much better than usual and I enjoyed it but I didn't hear nearly all of it as my mind was otherwise occupied. For one thing I was missing Scotty. He has gone up the line. I'll probably have a letter from him within a couple of weeks. There is a probability that I'll even have several letters following and then I'll never hear from nor see him again. So pass our friendships here! Friendships spring up quickly here and flourish for just a few days—the men are always awaiting orders here—and gradually dwindle away. Of course we get to know each other quite intimately if we are at all congenial because each day may be the last and it is hard to refuse to see them when we know how hungry they are for some one feminine and decent and of their own race and language. There is only a slight possibility of them being in this neighborhood again so when we part it is final.

I thot I had explained all this before but perhaps I haven't. I almost thot I detected a small green eye in your demand to know more about this fellow!! And I trust this will explain the situation—altho no explanation is necessary. There isn't a great deal of personal feeling in such friendships—of course there are exceptions. It is usually a matter of whom you happen to meet.

I'm afraid I'm not making my ideas very clear. I'd best stop and take another try at it tomorrow.

Sunday P.M. [***18th***]

I've been out witnessing a ball game between a Canadian team and an American one. The Canucks won but it was very close.

You see to what I am reduced for entertainment—ball games, poor concerts and that is all.

Last summer and in fact all winter I went to no ball games, attended very few concerts, didn't recognize the existence of men (with the exception of the one across the Atlantic) and hermited (is there such a word?) as much as is possible in a camp. As a result my roommate frazzled my nerves & I became as crabbed and unhappy that I just naturally had to come out of it.

I now grasp at every scrap of entertainment offered and my disposition has improved considerably and I'm relatively happier. The entertainment is often far from entertaining or even amusing but at least it is not hospital.

I've just thot of a way to get this thru. I'll give it to Conny and she'll give it to some one to frank. Having once found a way to get my correspondence thru to you without a third person seeing it I hate to come back to being censored. Anyway I don't know anything of military importance even had I a desire to impart information.

There was an air raid in Paris the night we were there but to us it seemed very tame. We didn't even hear the anti aircraft guns.

All has been quiet here since we returned. Today is clear but windy and there will be a fair sized moon tonight. Come over and sit on a sand dune with me and watch the fireworks [air raids] or perhaps merely the moonlight on the water. I'd love to have you.

On the way from Paris our train pulled beside a train load of Americans. We leaned out the door and called "Hello boys!" and you should have heard the yell they gave! And such a talk fest! They had just been over a month and were on their way up the line and they were apparently quite pleased to see some American women. And we were just as tickled to see the Yanks. I think everyone on our train had his head out to see what all the commotion was about.

Well, Buddy Boy, I must trot out into the ward and get an arm bath started and get our five bed patients fixed up for the night. . . .

Letters # 103 and 104, Laura to Erle, were brief and personal, the first undated and the second, August 26, 1918.

Letter # 102

Laura to Erle
Aug. 19, 1918
915 P.M.

Dear Buddy: —

The Ladies Home Journal advises against writing discouraged letters to our boys "over there". There is much suggestion as to what not to write and very little of what to write and there is absolutely no advice given in anything as to the proper tone of voice in which a red cross nurse should address her soldier love even when they are both over there. So if I err please lay the blame on the magazines lack and not on me. . . .

Have you made any other French friends? And how do you like them all? And don't you sort of like the funny little cottages with the red tile roofs? But perhaps, or rather probably, architecture differs in that part of the country. Are you in Normandy? I think you are but my idea of those old divisions is rather vague. I know this is Picardy and that Chamonix is in Savoy and they are mostly different provinces and different peoples.

Oh! I'm so disappointed that I didn't come down to Nantes! I know I could have if I'd only tried but it might have turned out as did another case here. One of the nurses went to St. Nazaire to see a friend. She saw him once for an hour and a half. Of course there is a possibility that he didn't make much of an effort but camp rulings are usually rather strict and unexpected things turn up. . . .

Did I tell you that the ward I'm in has a number of Yanks in it? I do enjoy having them. [***Laura later related how once she, a Hoosier, had to translate between a Virginian and a Yorkshireman who were in adjoining beds but could not understand one another.***] . . .

I'm in one of our three aristocratic wards where there are but thirty beds and a kitchen and nurse room and pantry and a closet for the extra blankets and wash basins and our beds have white spreads. Perhaps all of your wards are built so. If they are you can't appreciate the conveniences.

Frances Hampton had a victrola arrive this week sent by her home town. It has been on the way just ten months. . . .

The Final Allied Push Begins

Pvt. Baker's memoir:

August 27

Was sent to the Operating Theater for duty through Capt. Krost's recommendations. I have been wanting this for some time. We see all kinds of wounds. Plans are set for formation of a Casualty Clearing Station team. Jardine and I are going to be Captain Krost's surgical assistants. We enlisted for this special duty. I have been waiting for a chance to see the real front.

August 28

British are pushing right on at a terrible cost. Every inch of ground is fortified—pill boxes everywhere. The Huns have constructed concrete and steel forts all along the front in the Allied sectors. These pill boxes are built down in the ground, only a few feet being exposed and are cleverly camouflaged which makes them hard to locate with artillery. Machine guns are used in them which result in a terrible slaughter of attacking troops.

There is a continuous stream of wounded and gassed men entering the hospitals as a tremendous drive on a ninety mile long front is on in an attempt to break the Hindenberg Line. There are about 1,500,000 Yanks in France now so the Allies are tearing loose. The 27th and 30th Divisions of Americans are up here with the British and the French. All the British Reserves are being rushed to Picardie [*French spelling*].

British are pushing right on at a terrible cost. Every inch of ground is fortified—pill boxes everywhere. The Huns have constructed concrete and steel forts all along the front in the Allied sectors... There is a continuous stream of wounded and gassed men entering the hospitals as a tremendous drive on a ninety mile long front is on in an attempt to break the Hindenberg Line.

Letter # 105

Laura to Erle

"Rec'd at Nantes, France Sept 12- 1918"—By this time, mail delivery was reasonably normal, although Erle had yet to receive several letters Laura had sent in June and July to his business address in Chicago.

Sept. 5, 1918.

Dearest Buddy: —

. . . Nothing much has happened. We, too, are busy and I haven't had a great deal of time to myself. We are not getting terribly heavy casualties but a lot of them. Our hospital is filled almost continually and convoys and evacuations daily—sometimes three and four a day. And when I come off duty I haven't as much energy as I'd have if it wasn't such a rush. But just the same the work is nothing like it was last fall.

Isn't the news good? I can hardly wait for the paper to come to see what has happened. There hasn't been much in about the Americans recently but I suppose they'll soon start to push also.

And you are now an official evacuator. [***Reference to Erle being assigned to the detail which met ambulance trains, loaded the wounded and sick Americans into Dodge ambulances, and took them to the nearby base hospital. Later he was put in charge of the unit.***] I'm so glad you find the work agreeable even if it keeps you very busy for more hours a day than it should. . . .

Last Saturday the officers gave a little dance and we had a fairly good time—just about the average. I had a guest, Scotty had come down from his base about thirty miles away and it was a means of entertainment. He has left his base for the line now. He surely did enjoy the dance. . . .

Are you studying French? I'm struggling along with it. I've taken a couple of lessons and think I'll keep it up a little while. It keeps me out of mischief. The only trouble is that I have no use for it here at all. I wish we had French patients and then maybe I'd learn to speak it.

We have a number of New Yorkers in the ward. They all say foist, goil, shoit, etc. and I have to laugh at them. I've never heard anyone really talk that way before. I thot it was all in the magazine stories that they did so. . . .

Letter # 106

Laura to Erle
France
Sept. 8, 1918.

My Dear: —

. . . I'm glad to hear that your surroundings are fairly congenial as to the people you work with and that you have a job that requires a little brain work as well as physical exertion. I'm sorry your hours are so long—that's probably why you've been under the weather.

Our work has been increasing lately but today it was fairly light. Our worst case is an Australian who went out to the village last night and got all tanked up and then walked in front of a train and had two legs cut off. He had two pals with him and it took about fifty Englishmen to bring in the three. Then they refused to go to Hospital 20—in the same camp here—and demanded that they be brot to 18 "where the Yanks are." And finally they arrived. No one had put any tourniquets on the legs and the man had lost a fearful amount of blood. They transfused him but his condition is pretty poor.

Today he is sober and keeps wondering what "his girl will say." Isn't it too bad? They are such wonderful fighters and he might have lost his legs in a much more honorable fashion.

That's the only excitement we've had in camp for a long time. We haven't even had an air raid signal for some time. "Jerry" probably too busy to spare planes for bombing this far behind the lines.

To-day we had ice cream in our ward. The cook in the sergeants' mess made it and brot it down to us. Then this afternoon I got some sugar from the same source—most of our sergeants are too sleek (like your "Yellow Belly" of whom you wrote)—and made the boys some fudge. There was just enough for two or three bites each but they did enjoy it. They acted like a bunch of kids over it. We have two Australians, one Canadian, four Americans, four Jocks and three Englishmen in the ward. They are a nice bunch of patients.

Tomorrow we'll probably be filled up again but just now we are slack.

Do you have an American canteen in your camp or within reach? We have had one twice for two days. That is as long as the supplies have lasted. There is one now open for two hours in the morning but there is very little we can buy. I bought some more cigarettes—probably you'll get another package some day if I don't give them away before I get around to wrapping them up. If you don't care for the brand just say so and I'll not repeat the offense. . . .

Monday Evening

Well, my dear, I've just come in from dinner in the next village from ours. We had an excellent meal in a very nice French home that has recently started to serve to outsiders. It was sprinkling when we came out and it rained in our faces all the way home—about a mile. Now there is just a comfortable patter on the roof and I'd like very much to curl up in a big chair by an open fire and exchange a few confidences with you.

That seems hardly possible tonight so I'll have to get along as best I can.

It seems such an outrageously long time since I've seen you that I wonder if we'll recognize each other when we come fact to face. You remember that I have red hair and a high forehead and am not very tall, don't you? [***Laura stood 4'11' tall***] I'd hate you to be disappointed when you see me.

I remember you distinctly as a short, fat man with beetling black eyebrows and an iron gray moustache. [***Erle stood 5'11", was slender (165 lbs.), fair complexioned and sandy-haired.***]

I've had very little mail from home lately and all these clippings are quite ancient [***only one found in envelope***] but anyway they'll interest you. I don't think I know a man in this last draft.

My roommate is on night duty and it is such a relief to me to be alone in my room evenings. . . .

Letter # 107

Laura to Erle
France
Friday Evening, Sept. 13, 1918.
10 o'clock

My dear Mr. Davis: —

. . . I've just come from the nicest walk and you'd have enjoyed it very much I'm sure. The moon isn't very large and the clouds are quite black except just to the south west where the moon shows thru occasionally. The wind is blowing and the trees are swaying about. When we came back into camp we could see seven or eight search lights and their effect on the lower clouds was wonderful. We stood out and watched them for a long time.

Then we came in and read the "Stars and Stripes", the "Tribune" and the "New York Herald" about the American push. The first days gains are evidently satisfactory.

I can hardly wait from day to day for the paper. Most of the time we get only the "Daily Mail" or perhaps an "Express" because the American papers are very irregular. But strange to relate on the first day of the first American offensive we got all the news obtainable.

Have I told you, Honey, how tired I am of almost everyone that I know here? The one exception is Gabrielle Horowitz. She has had such an interesting existence previous to her emigration to the States that she fascinates me. Tonight she talked about one of the boarding schools she had attended in Marseilles and then about the various careers of the various classmates she had had. . . .

Letter # 108 was a short message with neither salutation nor usual signature, dated 15-9-18.

Pvt. Baker's memoir:

September 17

Left Camiers at 6:30 a.m. in a Buick ambulance for a Casualty Clearing Station at the front with surgical team composed of Captain G. N. Krost, Mrs. Krost—helper, Miss Silcox—anesthetist, R. Jardine and George Baker, 1st Class Surgical Assistants and John Goonery, Orderly.

We went through Amiens and then up through Albert. Albert has been completely pulverized by large shells as it has been lying right between the English and German lines for four years. In this town was a high church tower and about half way up the tower, there was a statue of Christ in a sort of alcove. The structure had been shelled and the statue leaned out on a threatening angle for a year or more until an English wireman guyed it up. Although the town was being constantly shelled, the statue came through it all safely until the fourth year when it was hit. It became quite a mystery to the British troops. Nothing over eight feet high was left in this town. The road was full of rocks, tile, brush, shells, wheels and all the waste of war.

We wound our way, snail-like, through all these places. The farther we went, the worse the waste. There were great holes in the ground as far as one could see in every direction. Many little groups of white crosses were to be seen where the dead from both sides were buried. There were trenches, barbed wire, ruined guns and shells everywhere.

We drove over what was once a perfect road but which was now full of shell holes to Paronne. Albert, Paronne, etc., are names as familiar to us as our home towns as so many of the wounded that we have taken care of were injured at these points. The Huns were on this ground only three days ago,

We finally reached our destination, a place called Brie. There is nothing left of it at all—only a few road signs telling us what it once was. We stopped at a few tents which had been put up on the muddy ground. We had driven over seventy miles and were now at C.C.S. #48, about seven to ten miles from the first line trenches. We had been twelve hours on the road. We slept in a large marquee with about fifteen other medical and surgical men. There were stretchers and blankets by the hundreds, ready for action. We knew a drive was going to start in the morning. Everyone was serious and there was no fooling—a taut, serious atmosphere prevailed.

I got my helmet and gas mask and about 9:00 p.m. we all retired. I had no more than laid down than boom-m-m went a bomb and then the z-z-z.z-z-z of the motors of the German night raiders. Everyone flopped to the ground. Our Archie defenders were busy a few thousand yards off. There were two Huns out looking for troops and ammunition trains. Some one yelled to come out and look at something. I dashed out of the tent and watched one of the Hun planes come down in flames a thousand yards away. It was a beautiful sight but terrifying in its flashing streak of death. I knew now we were pretty close to the active front. We heard another plane a few miles off and soon he flew over the first one which was still burning and which was lighting up the whole area for a half mile around. He dropped a bomb on this scene and killed seven or eight and wounded about twenty-five others, mostly Australians, who were gathered around the burning plane. The two German pilots were burned to death. Some of the wounded were operated on at our station by the night crew. All that night I slept with an ear cocked, nervously awaiting what might happen next.

September 18

Zero hour was at dawn when a thousand guns of all caliber were turned loose and the ground shook from the concussion.

We were on duty from eight in the morning until eight at night. The wounded started coming in on stretchers by six o'clock, or an hour after zero hour.

This is certainly a perfectly organized war. Here we get all set, then comes zero hour and a few thousand become killed, maimed or gassed. We are now prepared for anything. Our bridges are cut off behind us. We are over thirty miles from the nearest water supply. We are ready for the worst. Everyone is very nervous.

We operated right through for twelve hours with only a few minutes for tea, jam and bread for lunch.

Word comes from the line that our barrage was terrible, and that our troops had advanced a few kilos.

The wounded are carried back a thousand yards or more by prisoners under guard and are then placed in ambulances, if they can get to them, otherwise, four prisoners carry a stretcher all the way on their shoulders.

During the few minutes I got to glimpse at the sky during the day, I saw several captive German sausage balloons, as well as many of ours, all along the battle front, patiently watching the progress of the fighting. There were a lot of battle planes tearing around and ground strafing—a lot of action.

The wounded we handle here are nearly all in bad shape, with large pieces of shrapnel in them. We had to take off several arms and legs today.

It is all speed with us. We don't know whether we are at war or are dreaming. We are just about crazy. Everything demands speed, speed! If I hadn't had a years' training for this at the hospital at Camiers, I couldn't stand all this bloody work. I have two pairs of artillery leather lined leggings and two pairs of hobnailed boots. I have to scrub the blood off the leggings with a brush and soapy water. There is a creek down a hundred yards from our tents and I get water for washing there. It sure was a bloody creek just three days ago. The Germans had just been driven off this ground a day or two before we came here.

September 19

Very busy in the operation rooms. We have five other British surgeons and staff in our "Operating Tent". The wounded are lying in rows all around the large marquee. Most of the poor devils are asking for water and cigarettes—two things almost every wounded man craves.

It is a grand sight to see the great flocks of our planes going over the enemy line every morning between seven and eight for observation and bombing.

We operated from 8:00 a.m. until 2:20 a.m. the next morning—straight through, eighteen hours of steady work. Then we got up at 7:00 a.m. for another day's work. It sure is tough on us. We have some heavy cases now during this big Cambrai affair. I certainly feel sorry for the poor fellows who lose an arm or leg. They never can get them back. If they were in any other place but a lousy, bacteria-infected front, a great many of them would not suffer these losses, but here infection starts as soon as a man has been hit and it works fast. We have gas gangrene [*infection, not poison gas*] in nearly every case.

When I went to bed at 2:20 a.m. I was so darned tired that I wouldn't have gotten up for a bombing party.

This is the big Cambrai affair. We are advancing on a ten mile front and there are probably 200,000 Allied soldiers in this sector. The 30th Division of Americans are in it also.

September 20

We had another busy day. There was a great barrage again this morning at dawn. The ground seemed to roll from the continuous roar of the cannons, and the guns to groan with hatred. It is claimed that the cannon are nearly hub-to-hub in this terrible barrage. Usually shelling is in progress for half and hour before the troops start over the top, then the gunners lift the barrage a hundred yards at a time and our soldiers follow the barrage very slowly. The Huns either flee before this rain of shells or surrender when the Allied soldiers overtake them in their dugouts.

We are almost crazy with so many wounded coming in but if they can stick it out, we can. My legs and feet are what bother me the most. They get almost paralyzed from standing up so long. I have to bathe them in salt water every day when I get through. [*Yet Baker's fine athletic physique stood him well in all this labor, as it did the many other college athletes who'd enlisted in Unit 12.*]

At night the sky is lit up with the various lights, rockets, etc., and the line can be located as far as one can see. The poor doughboys have a lot of guts to stick it out.

Everyone believes victory is only a matter of time now.

Letter # 109

Laura to Erle
Picardy
Sept. 21, 1918.

Mon Cher Ami ***[my beloved friend]***: —

. . . Are you in a sandy country or muddy? It is very sandy here in camp but of course we are in the old sea bed.

Conny is down at 46 General quite sick. I went down to see her yesterday. She doesn't look very well but I hope she'll be better soon.

We've had such a lot of sickness among our nurses recently. One is not well enough to be sent home and probably never will be and another—one of the best liked in the whole outfit—has recently had a pulmonary hemorrhage and will go back to the states as soon as she is able to travel.

I do wish we'd move back of the American lines. I'm all tired of Englishmen and their ways.

There's certainly much difference between them and the colonials and also the Scotch. But I'm sure I've told you that before. Forgive me if I repeat. You see I've written you frequently in the past sixteen months and yet nothing new occurs.

Miss Horowitz of whom I've told you frequently has just had a letter from a friend of hers in Unit 11. The friends name is Miss Jones and she is supposed to be small and red headed. Do you know her?

Who is the chief nurse? I don't think I know any one there but there is a possibility that I do.

Sometimes I think I'll ask to transfer to a Mobile hospital unit. But now I'm going to wait till after I have another leave. I'm much more apt to get my leave here when I ask for it than if I am transferred among strangers.

By the way my little English artist friend has gone back to England.. The German drive toward Amiens in March made her very uncomfortable and she was uneasy all the time.

I've had some delightful afternoons in her cottage.

Are you studying French? But of course you don't have time. Unless you may be thrown with the French people a great deal you'll probably not take much interest in the language.

Did I tell you that Conny does very well at it? She has quite a remarkable memory. And then she has worked hard at it while I with my usual dolmens have wasted good time at it.

Do you ever get the New York Herald (Paris edition of course)? The Sept. 19th edition had a note in the Saint Mihiel write up about what Lt. V. A. Terrill (Francis Dobbin's husband) of Franklin, Ind. had to say about things. I'm sending the only copy I could get on home to the folks. I thot perhaps you'd see it anyway. . . .

I had a nice letter from Bob Murray today. He's still up the line and very busy. If he comes back to the rear base he'll come down again to see us.

Today a young English officer came to the ward to see our patient—that Australian of whom I told you and he came into the bunk (meaning office) and sat down and told me all about his wonderful exploits and what a detective he is. I felt as if I were speaking to Sherlock himself.

He even asked me to go to dinner with him and offered to read my palm. All of which I was forced to ignore—never having seen the man (pardon me, officer) before and having not the faintest idea what his name was nor did he know mine. But perhaps his instinct had given him my name and address and past, present and future state.

Of course they aren't all like that but honestly I've met so few English officers with anything interesting about them. And the conceit of them!!! They are simply unable to imagine anything civilized that isn't on their little old island.

You know, my deah, I don't always feel so about them. It just comes in waves and just now I'm having one over that ridiculous 2nd Lieut. . . .

Private Baker's memoir:

Casualty Clearing Station #30 near Ficheux, France

September 26, 1918

We left Camiers at 7:30 yesterday morning and had three ambulances for two teams. We rode through St. Pol, Arras and south to Ficheux, seven kilos, southeast to C.C.S. #30. We could see our captive balloons and some of Heinie's about fifteen kilos off. We got settled the first night about 7:30 and it was dark soon after.

The planes began to buzz everywhere and we could hear the bombs hitting up toward the line. Shells were going over our heads into Arras, five miles back every few minutes and as we lay in our bunks we listened to them sing and then burst with a big boom. We are in the center of large salient about fifteen miles across with the line on three sides of us. Fritz bombed all around during the night. At 4:30 a.m. the barrage opened and we knew our boys were going over. There were a couple thousand guns easily. Had lots of wounded and worked until 1:00 a.m. No sleep.

September 28

Very busy. Many terrible cases. Some men had seven and eight different wounds, mostly shrapnel. We had to operate on a couple Huns who had been wounded. We had a few hours off and walked around the battle fields which were everywhere. We encountered ruins, burial parties, wrecked tanks, dead horses, etc. I picked up a few souvenirs. Jerry was shelling all the area behind us and especially Arras.

September 29

Worked until 2:00 a.m. on very severe cases right off the great Cambrai battlefield. Very tired. Many shells and bombs were dropped in our area which made our work a nerveracking experience. We were too busy to get any souvenirs. We had ten or fifteen minutes for eats. I am in good with the cook so he holds out food for us. We just swallow it half the time.

Very heavy casualties coming in all the time. Operating theater is like a mad house and no one has any patience. We sure do see some wounds and the Canadians certainly are men. We have only a few wounded Huns in waiting now. Most of them have lost an arm or leg and Hun officers ask for American surgeons for them. They do not trust the English surgeons. We take off arms and legs almost continuously during these drives to prevent gas gangrene and b. tetinii [tetanus?] from spreading. It is a sickening job and the victim always has our deep sympathy.

Letter # 110

Laura to Erle
Sunday, Sept. 29.

Dear Man: —

. . . Are you busy?? We are up to our necks but not nearly as severe cases as we had either in the spring nor a year ago this time. . . .

It is cold and rainy tonight and my roommate went to church like a good girl but I just feel like visiting with you. I wonder if you are out evacuating and convoying tonight?? We've been getting convoys all day and are expecting many more tonight. We've been getting wounded from two parts of the line but most have been walkers. . . .

Well, what do you think of the war? I am so enthralled I can hardly wait for the papers and the communiqués which we get very seldom. Today there were big headlines but little real news. I hope Cambrai is ours by tonight but it may hold out a while yet. General Foch is my idea of a general. He seems to know his business don't you think so? . . .

Monday

. . . All day we've been getting convoys. Our place is full to overflowing. In some wards they have patients on the floors. But still they aren't such heavy cases as last Fall.

I didn't see a paper today so after supper I went back to one of the wards to borrow a paper and see what is happening on the front. I'm so excited about it all that my head spins most of the time. I've heard tonight that Bulgaria has surrendered unconditionally but as I haven't seen a communiqué I don't know. I only hope that Turkey and Austria will follow her example if it is true. . . .

Private Baker's memoir:

September 30

Very heavy casualties coming in all the time. Operating theater is like a mad house and no one has any patience. We sure do see some wounds and the Canadians certainly are men. We have only a few wounded Huns in waiting now. Most of them have lost an arm or leg and Hun officers ask for American surgeons for them. They do not trust the English surgeons. We take off arms and legs almost continuously during these drives to prevent gas gangrene and b. tetinii [*tetanus?*] from spreading. It is a sickening job and the victim always has our deep sympathy.

October 1

Still going strong. Long hours and many wounded. At this station we turn out eight or nine operations a day. Our seven surgeons and assistants total around seventy-five to eighty cases in twelve hours. Most of the casualties are Canadians and every third or fourth man of them lives in the U.S. so we are proud of them. They sure are glad they have fallen into American hands. When an arm or leg is to be taken off, we have to notify the poor patient why it is necessary to do, so he cannot hold the Government responsible after the war. It is a terrible thing to tell them but they have nerve and when they see it is the only way to save their lives, they muster up all their courage and say, "All right, Captain". I tell you that is fortitude.

October 2

Early and late, many cases waiting and the tables are full all the time.

Fritz dropped a few shells near our show house and a whole lot more a few miles back of us. We all stop work when a shell bursts nearby and all give sighs of relief when it is over and we find ourselves still standing by our poor wounded. It is dreadful for these poor fellows to get hit, then have to lie down and be operated on under ether a few hours afterward, right under shell fire. Hell surely can be no worse than this.

We have many Chinese coolies for road repair.

October 3

Busy all morning but our team was off in the afternoon and R. Jardine and I piked off for a long hike over the battlefield southeast of Arras. We were well satisfied that things were quiet while we were scouting around. We kept on the lookout for Hun airmen for they have wicked machine guns and use them at the slightest provocation.

October 4

Worked all day again but the cases were finished and we got to bed early. It is beginning to get cold and we feel it keenly. The three Canadians—Banks, King and Robertson—are in our tent. Went out souvenir hunting at a dump and got a lot of stuff.

I know if it were not for the darn cooties [*lice—known disease-spreaders, in addition causing itching and bites*] that keep us busy scratching and digging, that we would go crazy. It keeps our minds off of being scared to death and from our primitive living. They multiply about 10,000 times in twenty-four hours. They seek the warmest spots on one's body, like the neck band, under the arms and legs, and Wow! how they do eat. We have to take off our underclothes and pick the cooties off one at a time and each has his private funeral between two thumbnails. This is the only practical method found to kill the little pests. Each wounded man has an army of little friends on him and we get them also. Our blankets are wonderful nests.

October 6

At 10:30 a.m. we left the front for our base and made it home at 4:00 p.m. We stopped in Hesdin for eats. I bought Sis a pillow top for her birthday. Gave Sid Barteau one of my big shell cases.

I am back in the Operating Theater and expect a call again soon for the front. Got a bath and feel fine though rather spent from the long tedious, anxious hours and the loss of sleep.

October 7

Very busy in the Operating Theater. I worked until 5:30 when all the cases were done. Am handling same kind of cases we started on at the front. Some of the poor fellows got worse and had to be operated on again. Others, and a lot of them, went West. Poor devils. It is next to impossible to prevent infection.

October 8

Lt. Chamberlin, of our Aviation Service, came over and pulled off a lot of stunts and then landed on a hill near our hospital and came into the Theater for a dressing on his face. He is a quick, keen, spontaneous-acting fellow and is about 5' 9" in height. We went over to look at his new "Sopwith Biplane" which he brought over from England to France in twelve minutes. It had a 200hp Nome motor and hits 140 miles an hour. If we had beaucoup [*many*] of these planes, Fritz would soon be crowned.

October 9

Very busy in the hospitals. Cambrai drive is in full swing again. Operating Theater doing around 75 to 100 operations in twenty-four hours.

Fall 1918 and Armistice, October-November 11

RUMORS AMIDST OVERFLOWING WARDS

Letter # 111

Laura to Erle
"Rec'd at Nantes Oct 19-1918" (Erle's 31st birthday)
Oct. 10 1918
130 A.M.

Dearest Buddy: —

Now I've three letters to answer one of Sept 20, one of Sept 25 and the last and fullest Oct 1. And under the present circumstances letters are almost a physical impossibility. I went on night duty a week ago Mon. or Tuesday I forget which. With the push on and the hospital running over its capacity and having the two heaviest wards in the place I've just been unable to see over the stacks of work. Last night and tonight I've had just one ward and it has been a great relief to me. I couldn't possibly do justice to all the sick patients I had.

By the time I was off duty I was so tired that I could scarcely navigate and to think or write was simply impossible. I received the first two of your letters the first night and was so very glad to get them. And the last one containing the post card views of Nantes and vicinity came yesterday.

Nantes must be a very pretty place. I'd like so much to see it. And it is certainly very fine for you boys to have met some such pleasant people as M. Schwab (that name sounds Teutonic) and the two French girls. It will help to make things pleasant for you and the camp life more endurable.

Yes I know Gunhilde Johnson. I always liked her very much altho I didn't know her at all intimately. There's something so substantial in her characteristics. Don't you think she is rather good looking? I do. Is Ingrid Johnson another Augustana nurse in your outfit? There may be several that I know.

Last night one of my men hemorrhaged five times while I was on duty and the night before another little Canadian had a hemorrhage from the femoral artery so you can imagine what close watching I must do. We've had almost no light cases in these wards this week and a half. They are very heavy. Most of them are Canadians but a few Imperials and a very few Americans. . . .

Evening

I do hope you get your sergeantcy. I feel sure that was what you meant by what you hoped to be able to tell me.

It hasn't been nearly so cold here as it was last year or else I'm becoming acclimated. Anyway I haven't suffered with cold so far and have developed no chill blains.

I'm very nearly a third thru my night duty. I'm not sorry that so much is gone. A month isn't so very long after all, is it? But a year and a half is ages and eons (?sp) when I can't see my sweetheart. I hope it won't be much longer.

Private Baker's memoir:

October 10

Long hours and we are running at top speed. I hope we get a rest soon for it is a killing grind. I can't get caught up. We work twelve to eighteen hours. Three times a week the teams work the extra six hours in relay.

October 11

Still going strong and no end in sight. It has been an exciting time in the whole hospital for we have more patients than beds. All we do is carry wounded in and carry other wounded, bound for England, out.

All the men are working so hard here. Last night they carried stretchers all night. And of course the day work is to be done as usual. Some of them bear up beautifully under it and others fall down miserably on every job they have. Funny world isn't it?

Do most of your convoys and evacuations come at night? Ours do almost always.

You spoke of a Mencken review you were enclosing. Well you didn't enclose it but I liked the little South Wind verses by Hortense Flexner. She writes beautifully.

Did you see in the paper that Mr. Patterson [***Erle's former boss***] is in Paris? I suppose you knew about it.

What does Hib (pardon the familiarity. I'm merely quoting you) do in civil life? Is he a newspaper man too? And what became of the man whose sweetheart was in China and you helped him write a conciliatory note?

It is so nice that you are near a fair sized town where there are shops and cathedrals and even a theater. You should see our town. It consists of one street so [***drawing of curved road***] and it is a filthy place but rather picturesque. . . .

October 13 and 14

Big hurry to get everything ready. I was not quite ready to start and got bawled out so shaved while riding in the back of our ambulance while going 35 miles an hour—some trick [*with a straight razor?*].

We went to St. Omer first, then to Ypres, Belgium, where we put up at a C.C.S. Never saw so much mud in my life.

Fritz started shelling at 4:30 and kept it up all night. The Germans are on three sides and we can follow the brilliant lights, rockets, etc., all around us. We sure are in a hot place. We can hear their cannons boom after the shells break, and although we are in a direct line between the cannons and Ypres, the shells all burst short of us. First we hear the sizzling of the shell, then the timer explodes and finally the shell proper. The whole countryside resounds with echoes. When the first one came I knew Ypers was being shelled. We kept low all night. All of the roads in around Ypres were shelled all day but we were lucky for none hit less than 400 yards away.

Some doings last night. A Fritz came over and shut his engine off right above us, then he came down and machine-gunned the road and dropped a lot of bombs around the neighborhood. An American searchlight picked him up and our own boys brought him down with a machine gun.

Letter # 112

Laura to Erle
18 General
Oct. 13 1918.

Dearest Buddy Boy: —

I'm all excited over the news tonight. . . .

Do you think we'll have peace soon. Or is it all bluff or are the Allies going to wade right thru Germany? And are we going to be back home by the Springtime. No, I know you aren't an oracle nor a prophet nor even a mystic but I'd just like to hear what you think about everything.

Please excuse the smudges on this paper. The stove just blew up and spattered soot over everything and this my one and only sheet of paper. . . .

Today, when I got up I found a note waiting for me from Robt. Murray. He has been wounded and is in the officers hospital in this camp. Frances and I went down to see him after tea. He seemed awfully glad to see us and didn't appear to be in bad condition. It is a deep flesh wound I think from what he told us. Anyway he needed a shave very badly and was out of tobacco and matches. I anticipate that the next time we go he'll be better because we immediately went out and got him some cigarettes and cigars and some matches. And I heard the orderly tell him he'd shave him this evening.

Today was our guest day at tea. It comes just once a month. We had coffee, sandwiches and doughnuts. The latter were delicious.

I've had several letters from home this week containing all the home news but not many clippings. You know, my dear, I don't crave American mail half as much as I used to before you came across. Of course I want to hear from home but it is different from hearing from you. Now I await the local mail with much interest.

Are you so fearfully busy as you were? We are but as I have just one ward now and an orderly I get on nicely. Just now I've only one very sick man and he is sick enough and uncomfortable enough for a ward full. He is a Canadian and is surely patient. I very nearly special him at night. . . .

Dining hall set up for a celebration.

Private Baker's memoir:

October 15 and 16

Water and mud up to our ears and it is impossible to walk for shell holes and mud.

Some doings last night. A Fritz came over and shut his engine off right above us, then he came down and machine-gunned the road and dropped a lot of bombs around the neighborhood. An American searchlight picked him up and our own boys brought him down with a machine gun.

We were very busy in the [Operating] Theater. The boys are advancing very fast all along the line. Menin, which is straight east of Ypres, has been in our hands since the 14th. Lots of civilians are coming through the lines, many having been gassed or shelled. The next table operated on a pretty little Belgian girl of fifteen, a pitiful sight. She had shrapnel in her ankle, which necessitated the amputation of her foot. We were all upset over it.

I am doing this little scribbling in my tent by the light of a candle covered by a Hun helmet.

It rains about every day. I keep a pair of British boots hanging up in the tent drying while wear the other pair. The mud holes are terrible.

October 17

Started work at midnight when we received some wounded. The Chinese coolies carry our wounded around and put them in ambulances for us. The British brought 200,000 of them over here to do labor work.

Letter # 113

Laura to Erle
Oct. 17, 1918.

Dear Honey Boy: —

. . . I've just two more weeks of this night duty but if it were all as calm as tonight, I shouldn't mind in the least. In fact unless a convoy appears before morning it will become quite stupid. I've nothing to read and it is impossible for me to wax enthusiastic over letters at this hour of the day. Anyway there is nothing to write about. One day is just like another. . . .

Tuesday

2 A.M.

How I wish things could be completely finished this Fall! But the good weather won't last much longer and necessarily they can't do much pushing in the winter.

Are you managing to keep warm? I've been fairly comfortable most of the time so far and my night duty will be over in another two and a half weeks. That seems a longtime when I dislike it so but after all it passes quickly.

I'm going to try not to hibernate this winter as I did last. But it is rather difficult not to when the nights are so long and there is no place to go. Perhaps another year will find the world at peace and we will all be more happily situated.

Have you heard of your father's condition recently? Has he improved? You haven't mentioned it for a least a month.

In this letter there is the first mention of the deadly influenza epidemic that swept the world in 1918. A mild form of the flu was thought by some to have contributed to the failure of the German push in the spring of 1918, when soldiers on both sides were affected. The deadly form of the disease began hitting in late August 1918. It is amazing, given the rapidly spreading characteristics of Spanish Flu, that it had not reached either Base Hospital 12 near Flanders nor Base Hospital 11 at Nantes before late October. Ultimately, the 1918-19 flu epidemic was blamed for as many as 50 million premature deaths.

Influenza Epidemic

Letter # 114

Laura to Erle
Oct. 19, 1918.
[No salutation]

Birthday Greetings!!!

. . . I came back past our village and purchased grapes and tomatoes for some of the boys in the ward and a Daily Mail for myself. And now here I am with my ward nicely settled and all ready for the night. There are a few gas cases wriggling and one or two sitting up looking about but almost all the men are asleep.

So have I spent your birthday. What kind of celebration did you have?. . .

I've been so sure for the past two days that I'd have a letter from you. I hope you haven't developed Influenza. I haven't heard from you since Oct. 1 and so many cases of In[fluenza] have developed among the men in camp and apparently all over every place that naturally I've begun to worry about you. . . .

I've sometimes thot that the slowness of our mail was caused by us being with the B.E.F. and so having to pass thru two branches of service. But perhaps it is all like it is here. . . .

Do you know that for a year and a half I've been writing you two letters a week approximately and that is about one hundred and fifty? [***This would indicate that quite a few may have been lost since this is #114***] . . .

Early 20th century sterilization system.

Letter # 115

Laura to Erle
Oct 21
1 AM

My dear Sergeant: —

Congratulations upon the promotion. I'm so glad you have it and I know you've earned it. It is quite an encouragement, isn't it?

I've just received your letters of Oct. 3 and 12th about three or four hours ago. My roommate came over to the ward with them and I was surely pleased. It has been all of two weeks since I had your letter of Oct. 1 and I was getting very, very hungry for more. . . .

I seem to know very few of the North Vernon people any more even when I have been there fairly regularly. But occasionally I see a familiar name. The draft lists are practically all men that I've never heard of or else they are boys that used to be my pupils when I was a country school marm.

Today I had another letter from Bertha—written Sept. 29, four days before yours—and she tells me that John Green ***[Laura's cousin, a North Vernon physician]*** has had an operation (knee I think) and has now reported fit for duty and awaits his call to service.

Everyone seems to be here or else on their way.

The news is certainly gorgeous. The Kaiser's days grow shorter so rapidly that it does my heart good. I do hope he gets all that is coming his way soon. . . .

All leaves are stopped here for a while. But of course yours haven't commenced.

You must get your convoys in differently than we do. You speak of two in forty eight hours. We get in as many as five or six in twenty four hours when the push is on. But of course we evacuate rapidly to England.

I've changed wards again tonight. And right beside me is a man so delirious that he needs attention of 2 men. So if this sounds jerky you will understand the reason why.

My orderly is a Mexican from N. Mexico. He isn't very intelligent and only fairly willing. So he isn't a howling success as an assistant. Fortunately I have only one other sick man and he is asleep. . . .

This is the first letter addressed to "Sgt. John E. Davis," with "Oct 30 1918" on reverse.

Enclosed in the envelope were a number of newspaper clippings, including a memorial service at St. Mary's Roman Catholic Church, North Vernon, for a distant Huckleberry cousin, Matthias Hester "who gave his life in the service overseas," one of a local man wounded in action, another about two men being commissioned as lieutenants at Camp Taylor, Ky., near Louisville, and other purely social announcements.

Yesterday we had a convoy of American patients from Belgium. They came down in boxcars were thirty-six hours on the way and had nothing to eat. I was so disturbed over that that I couldn't sleep at all well last night. And today I understand that it is fairly common in the south of France to send our wounded down to base in that fashion. Have you seen anything of the sort? It is unheard of with the British and I don't understand why our people aren't as well prepared. If the people at home knew of such a state of affairs there would surely be a grand and glorious explosion?

Private Baker's memoir:

October 22

Worked all night at C.C.S. #36 under Major Lockwood and we sure had some work, too. Had some bad cases. Rumors are afloat that Germany and Austria are willing to quit.

October 24

Moved at 10:00 p.m. from C.C.S. #17 back to Armentieres, Belgium. Having nothing to do at present. After getting settled we slept all afternoon and night. The wounded are being sent to another station which is nearer to the railroad. Cannot get enough sleep. I don't think I will ever get rested even if I do get home. My nerves are O.K., but much overtaxed.

Letter # 116

Laura to Erle
Oct. 24, 1918.

Dear Sergeant:

. . . Did I tell you that I am now in the head ward? It is not very exciting but quite busy. For an orderly I have a Mexican from N. Mexico. He claims to be Spanish but he surely does look like the Mexicans and as if he had a fair share of Indian blood.

How important the Sergt. looks before your name. Does it make you feel all puffed up? I'm so very glad you have the promotion. . . .

Are you still working eighteen or twenty hours a day? That is too long to keep up for any length of time and I hope you are getting more rest now.

We are sending out just one patient from our ward tonight and unless something unexpected happens we'll have a quiet night. The place is full up and there is no room for more. I've already three sleeping on the floor and I've no more mattresses nor blankets.

Just six more nights and I'll begin once more to be a human being. It has seemed such a long, dragging month even tho I've been busy. I've never disliked it quite as much as I have this time. I hope before my train comes again I'll be finished with the war and all such troubles.

I'm beginning—no not beginning—to wonder how long it will be after the war before we'll be sent back home and mustered out. I hope not very long. I'm so anxious to get thru with it all and naturally I'm about one in a billion or so wishing the same thing. . . .

Private Baker's memoir:

October 26

Jardine and I went out walking, going through St. Andre, la Madeleine and as far as Lille, which was not open to troops as yet. We talked with civilians of the town who had been in the hands of the Huns for four years. They did not know we were Americans as they had been ignorant of all outside affairs all this time. They took the Australians for U. S. men as we wore big hats once.

October 29

(Lille) I talked with lots of civilians. They told us of the visits of the Kaiser, the Crown Prince, von Ludendorf and Hindenburg and of how the French officials were forced to shake hands with the Kaiser and of his many medals, etc. None knew I was an American until I told them. They thought we were Belgian or English with our darn little caps—a disgrace to our army.

[*Erle remarked in later years how much A.E.F soldiers hated the French-style "overseas caps" that replaced the U. S. Army's traditional pre-war Western-style broad-brimmed "campaign hats." Campaign hats, he explained, provided protection against rain and sun, which the overseas caps of which Private Baker wrote, did not. But overseas caps persist in some U.S. military uniforms into the 21st Century.*]

The Huns cut off and ruined the lighting and water systems of the town and did many other dirty tricks before they left. They also took about 2,000 valuable paintings from the town art gallery.

Letter #117

Laura to Erle
Nov. 4, 1918

Dear Man: -

. . . Your letter was written Oct. 21st and gave an account of your work and of your dinner party with Grunhilde Johnson. That was nice for you both. Do you find her companionable? As before stated, I don't know her at all well—but I like her appearance very much indeed.

The other Miss J. I mentioned is quite a different sort but a good scout and fond of a good time.

Have you read any of H.G. Wells' books? I've just finished his latest one, "Joan and Peter." It was very readable but more or less disturbing to me—perhaps because of my mood at the time I read it. He certainly wallops the mismanagement of the war and gives credit where credit is due.

Yesterday we had a convoy of American patients from Belgium. They came down in boxcars [***The infamous "40 et 8s" (40 men and 8 horses), French freight cars used to transport American troops—to their disgust***], were thirty-six hours on the way and had nothing to eat. I was so disturbed over that that I couldn't sleep at all well last night. And today I understand that it is fairly common in the south of France to send our wounded down to base in that fashion. Have you seen anything of the sort? It is unheard of with the British and I don't understand why our people aren't as well prepared. If the people at home knew of such a state of affairs there would surely be a grand and glorious explosion?

Now all this isn't exactly private or family matters, is it? But I just have to tell it to some one even if it goes no farther than the base censor.

Oh yes, I fell heir to three more green envelopes. They are very precious just now and hard for us to get. . . .

The news is surely good. I don't think hostilities will continue very long, do you? And I'm so anxious to get back to the States and start to really live.

Evening

Well, Belovedest, evening finds your sweetheart in a cheerful mood. I don't know why the change. It surely isn't the ward because two of my patients are much worse tonight and I don't think they'll be here to greet me in the morning. But the others are fine and such a pleasant lot of men, quite a lot of them are Americans and I do like Yanks—most of them anyway.

And after all my bubbling over the box cars, Capt. Bird came in tonight to tell me that it was a French ambulance train. That is the kind they have. But that isn't good enough for our men. They should have a real train for them.

And you showed Miss J. my photo and everything. Did she recognize it? And talked about me? Again my curiosity rises. What did you find to say about a person who never did anything out of the ordinary in her whole life? But it's a foolish question.

Today's papers announce the complete surrender of Austria. They seem to be falling fast. But still the old Kaiser hangs on by his teeth and refuses to see his finish. But he is getting awfully shakey.

The family has come home and there is entirely too much talking to continue this. . . .

Private Baker's memoir:

November 7

Rain, rain continuously, never allowing it to dry up a minute. It is just the limit and very hard on our nerves.

Very few wounded are coming in.

All kinds of men are dying every hour with the flu. They cannot get them buried fast enough. It is a shame to die just as the war is about ended. An ambulance is loaded at the front with flu stricken patients but before it can get to the train, just a few miles away, most of the men are dead. They are lying in rows, each covered with a blanket. I sure feel sorry for these war-torn Tommies who have to die from flu when the war is about over.

Letter # 118

Laura to Erle
Picardy
Nov. 7, 1918.

[*No salutation*]

. . . The air today is full of peace rumors and tonight we hear that the armistice was signed at 3 PM today. But we've nothing authentic yet. Won't it be wonderful when the peace really and truly comes? I can hardly imagine what it will be like to get back to civil life again. It will surely mean a great deal of readjustment for the men who have been out so long. . . .

Have you had much flu in your camp? We've had a great deal but no severe cases among the staff. We've had no let up on the work since the last of September. I think this is the longest stretch we've had with no lapses between pushes. When the wounded convoys aren't coming we get medical ones and so have kept very busy. I've gone to a medical ward and I'm not very keen about it. I like surgery heaps better but just the same the men need some one to look after them in the medical wards. We've eleven Yanks and most of them are very nice. One is from Indianapolis and another from New Albany [Indiana]. . . .

M'deah, you should see me out robbing the coal pile. We don't receive any fuel for our rooms till Nov. 15 and in the meantime we aren't supposed to have a fire. But we manage to get a bit here and there and are nice and comfy evenings when we are off duty.

We found an unguarded coal pile and after dark we sneak over, Polly, Bess and I, with our little baskets and come home with some nice fat lumps. It's a gay life in the army!

Tuesday night I went to a dinner and a dance at an officers' mess not a hundred miles from here and it was so funny. They have entertained bunches of the nurses at different times but I had never gone before and wasn't very keen this time but was just curious. So I went. They sent a lorry for us and seventeen of us went. We had a very nice dinner only all the men gobbled everything so fast it wasn't hard to see they had been away from real folks for a long time. Then we walked thru rain and mud about a half mile to the dance. I went with an awfully nice Canadian major from Edmonton, Alberta. (I seem to like the Canadians better than anyone else I meet out here).

The dance was funny, too. Their waltzes are so fast

and they don't dance as we do at all. Their waltzes are just like this [*drawing of continuous loop*] one continuous whirl and when you get so dizzy you don't know where you are. They reverse and take you in exactly the opposite way. Before the evening was over I could get around the room a time and a half—and that means a lot of circles—without staggering. And you should have seen me doing "The Lancers" which apparently is a great favorite. My feet weren't on the floor very much of the time. It was one mad romp—and very funny. I wish I could sketch. I'd love to show you some of the poses of my partner, a little, blonde Yorkshire Major.

It was quite an experience. I had meant to be only a spectator but that was impossible. . . .

Friday Evening [8th]

Well, the war isn't over yet. Such a rush as there is for the papers and then there isn't usually a great deal of news. I do wish we could bomb Berlin tonight. I'm sure that would have a wonderful effect on her delegates.

It is hard that the Influenza is taking so many of the men who have been thru so much and were just nearing the end. It is very sad, don't you think so? The Flu is surely treacherous.

The men have a scratchy old gramophone and are playing all sorts of records but some way it takes me back to 453 Belmont, "The Land of the Sky Blue Water" keeps running thru my head and I haven't heard it since I left Chicago. It is a pretty thing, is it not? . . .

Private Baker's memoir:

November 9

Raining yet. Nearly up tour knees in mud. We have very few wounded.

Thousands are dying of the terrible flu.

I change shoes and socks every half day in an attempt to keep my feet dry, while I have clothes hanging up in the tent to dry.

November 10

Little work. It gets very cold during the nights. I use four or five blankets—mostly lousy [filled with lice] ones.

We have a few more cases—one German officer and a very large Prussian Guard sergeant who has lost a leg.

Monday, November 11, 1918

Armistice a.m., at last!!

At 5:00 a.m. a long range shell dropped into a big group of South African artillery in front of us and killed half a dozen and wounded fifteen or so. I felt so sorry for the poor fellows. They had to get hit the last day and 8,000 miles from home. We operated on one young South African fellow who had eleven or more pieces of shrapnel in him.

NU's Jazz band, mentioned often in m.s. Very popular at the various military posts in the area.

As published in his daily column in the Louisville Herald-Post,
November 11, 1932
by John Erle Davis

AN ARMISTICE DAY STORY

The Armistice in World War I was declared at 11 a.m., November 11, 1918, 14 years ago today.

Armistice Day and its mist of fading memories—some rosy and gold, some gray and sad. Most people had great fun in American cities that day and night. But there were some places where the fun was muted.

The scene is an American base hospital camp near an ancient French city where the cathedral cornerstone bears the legend "Anno Domini 1421." The morning is mild and clear. Vague rumors of the war's end are in the air.

An American colonel from the Argonne drive, dragging around on crutches, amiably discusses the possibilities with gas-burned buck prlvates.

A half-blind, one-armed English-speaking prisoner of war hopes to God it's about all over. He explains, as tears fill his eyes, that he has not in three years seen his wife and young sons somewhere in East Prussia.

Comes 11 o'clock, and the soft breeze is laden with the musical booming of bells in distant church towers. Hobbling wounded...French, AmerIcans and Germans...are grinning and shaking hands, the enemy pathetically eager to be friendly.

Groups are singing—the Germans some measured, mellow, moving song about home; the French their rollicking "Madelon"; the Americans their foolish, stuttering, gay "B-b-beautiful K-k-Katie, beautiful Katie...at the k-k-kitchen door."

Excitement mounts. It is quiet in the half-empty, long wooden-hut wards. The men are speculating on how soon they will be back in the States. The sergeant major is besieged for passes. Everybody who has francs in his pocket wants to go to town and celebrate, even the hobblers.

The mail comes. The orderly hands a cablegram to the favorite lieutenant of the outfit. He stares at the contents, then says, "Well, I'll be damned. My mother's dead." He giggles a little hysterically*.

A sergeant speaks to him, "I heard, lieutenant...I'm terribly sorry," and offers his hand. The lieutenant grasps it, then hurries to his quarters. There will be no celebration for him this day.

A long detachment of convalescent soldiers in double file, out for exercise, clop-clops through nearby winding village streets chanting the ribald American Army classic "Mademoiselle from Armentieres."

Now it is night. Place Royale and Rue St. Pierre glitter with festoons of colored lights that have been hurriedly strung on building fronts and across streets in the afternoon. Huge, drab, muddy trucks one after another stop by the big fountain and let out uniformed Americans keen for a big time.

The streets are jammed from wall to wall with French and Americans... soldiers, sailors and civilians. And girls. Everywhere girls—kissing, squealing, romping, making roughhouse with the men. Mardi Gras without masquerading.

Hours of this. Then pairs of American military police come pushing through the crowds, stopping at intervals, blowing whistles, shouting:

"Tenshun, soldiers! All hospital men report to quarters at once! All passes cancelled. Trucks are waiting at Place Royale…Tenshun, Soldiers! All hospital men report…"

Later the trucks rumble back into the hollow square of the camp. All huts, offices and kitchens are brightly lighted. Things are stirring. In the big Receiving Hut, stoves roar and become red-bellied, shooting dashes of sparks into the darkness above the roof. There is a sprinkle of rain.

A sergeant checks the count of clean folded blankets dumped upon a rising pile. Medical officers come in, and nurses in gray dresses, long blue capes and little white caps—ready for business.

The mess sergeant starts the cooks to making cauldrons of coffee and soup, and heating barrels of water for baths.

Over in the village, a long hospital train backs quietly and smoothly in on a sidetrack. Presently the ambulances are coming into the hollow square, to be unloaded by husky litter bearers, then gunning off for another load, and another, and another.

It is a busy night. Three trains. The sprinkle of rain becomes a downpour.

The Receiving Hut is crowded with rows of litters…holding the morning's crop of slaughter up on the lines…muddy, filthy, bloody, lousy, hungry, tired miserable men. Officers move back and forth along the rows…looking at bandages, reading tags, calling ward assignments:

"Depressed skull fracture. Number three." "Mustard-gas contact, Number seventeen." "Compound fracture of right femur. Number nine." "Phosgene gas inhalation. Number six." And so they go.

And the litter-bearers whisk the wounded away to clean beds, fresh dressings, hot food, baths, rest, peace, sleep… for some the first and last Armistice Day.

Dawn comes. The rain stops. Raw wind whips the dripping trees. They groan.

In the little Dying Room in the prisoner-of-war hut, a blond youth, his skin the shade of putty, lies moaning and murmuring on a bed moans, "Mutter… Mutter…Mutter."

A comrade stands beside him, fingering a rosary, reciting the Lord's Prayer in German, "Vater Unser, der du bist in Himmel . . ."

Standing by the window, a medical officer loads his hypodermic. "Poor kid," he says. "Gas gangrene. Not a chance he'll ever see his mother again…"

Armistice Day, and its mist of fading memories.

* *This was Erle's literary license. Actually it was he who received notice on Armistice Day that his father, John McAlpin Davis, had died in Indianapolis at age 70. See his letter home, following.*

Erle to his mother

Sergeant John Erle Davis
USA Base Hospital No 11
Amer E F APO 767
Brittany

November 12 1918

Dear Mamma, —

So Dad has gone on the Long Journey. I trust beyond the telling with mere words that he is with the many he has loved, who have preceded him into The Fair Land. That he received my letter before he died, is a source of deep satisfaction to me. And I am glad for his sake and for yours that the end — the death itself — came quickly and quietly. For you and the girls, I realize, it means the finish of a long period of anxiety, of many trials, of the exercise of great patience — in short, of heavy cross-bearing. I have been very fortunate indeed — Dad to me now is a memory, one of a quaint, lovable, character, unblurred in the manner that your memory is blurred, to a degree. I fully appreciate the position you and the girls have been in for the last ten years.

Betsy's letter telling of the death came yesterday at noon, just after we received notice of the signing of the armistice by the Germans. Whistles were screeching all over the city and bells in a hundred church towers were giving tongue to the glad tidings. The noise was like that in American cities when the old year passes and the New Year comes in. In the circumstances I was unable to rise to the occasion. I went downtown to file travel orders in the evening and for a little while watched the crowds. The streets were packed. The air was full of confetti and song and happy shouting — it was like election night at home much magnified, and they're still at it.

But, as I said, it left me cold. My spirit was with you and I wanted more than anything in the world just then to be with you in the body. I returned to camp and with the others worked until 3 am unloading a hospital train. I was up again at 6 o'clock and have been on the job since.

We're all working hard — long hours under considerable pressure. When there is more time I will write more frequently and at greater length.

So Dad has gone on the Long Journey. I trust beyond the telling with mere words that he is with the many he has loved, who have preceded him into The Fair Land. That he received my letter before he died, is a source of deep satisfaction to me. And I am glad for his sake and for yours that the end — the death itself — came quickly and quietly.

Armistice as Anticlimax

Letter # 119

Laura to Erle
Nov. 12, 1918.

Dear Buddy Boy: —

"Le guerre finnis" [***la guerre finis***: the war is ended] and what else is there to write about? My mind is full of things to talk about to you and last night I did so want you. Doesn't it all leave a sort of numb feeling? I haven't fully realized what it means yet and suppose I won't for some time. . . .

Last Sunday night—you know before the armistice was signed—I was studying the map and I had my leave all planned for early next month. I was going to come down to Nantes for a couple of days so I'd be sure to see you. I was so afraid we'd miss each other if we planned to meet some place. I intended writing that very night to tell you all about it but decided to wait & see if we were to have peace.

Now, I suppose there'll be no leaves and we may be going back to the States within a couple of months. I don't want to go till you do but I realize that my wants have nothing to do with what we'll do.

But if we stay quite a while and if we get leaves you can expect to see me hanging about your neighborhood for a little visit. Will that be all right??

Do you think your unit will stay where you are till spring? Now, I want to hurry right home.

And yet our hospital is full. We have a number of badly wounded and many flus. Now, please don't get so tired out that you take flu and be so sick. This flu is treacherous as can be.

I imagine in the course of a month this place will be quite clear of surgical and even of medical cases and I can scarcely imagine that they'll keep this whole bunch of hospitals open all winter. Wouldn't it be fine if after they close this they would send us south & split us up among a number of Am. units and I'd go to Unit 11? Pipe dream?? You know I have lots wilder pipe dreams than that!! . . .

Wed. (Nov. 13)

I had an afternoon today and spent a rather unsatisfactory time trying to shop and to catch up on some little things I'd been putting off. Tonight the officers gave a dance. The music was gorgeous but —-. Some people do a lot of lopping [?] over one each other, don't they? That sounds bad coming from me but I was really disgusted tonight with some of the performance. . . .

Private Baker's memoir:

November 13

Last day at Casualty Clearing Station #48 and the front. Took a hike to Maretz and Clary with Jock Wilson of South Africa and got three German shell cases for home. I met a lot of U.S. Railway Engineers at work building trestles and new tracks.

After dinner I took a lorry to Landrecies and Maroilles which was toward the line. We got up to the point where the fighting had ceased on Armistice Day. I got a good seat in front with driver. Had time to eat at a U. S. Signal battalion's mess and returned late. There were hundreds of captured guns everywhere. A few bridges which had been blown up were being rebuilt. The boys were all coming out of the line happy at the thought of war being behind them and that they were safe. The Yanks are given credit for ending the war. It's great to be a Yank in France now.

November 1918-February 1919

PEACE AND LONGING

Letter # 120

Laura to Erle
Picardy Nov. 16

Dear One in Brittainy : —

After many fruitless minutes and perhaps hours trying to solve the riddle of our future I've decided that since I can't possibly know anything about it I'd better let it alone. Henceforth I refuse to worry. I'm believing that it will all solve itself in time and I'm hoping that everything will be all right for us. . . .

How I'd like to see you a while to have a nice visit and talk things over! But I don't know when that will be. If we stay long enough and I get a leave I'm going to see you or bust (I'm not at all keen to bust) but that is all to be decided in the future. We may evacuate here inside of two weeks or several months. Just now we aren't evacuating cases. I don't know why. It may be because of sweeping the channel for mines or it may be that the English hospitals are too full. So we are filled up here with no room for new cases.

A letter from Bertha today states that they have had one death from flu in N.V. and they have brot two bodies back from cantonment. The schools have been closed for several weeks.

Are you hearing regularly from home? I find that helps a lot in keeping up my morale. But that doesn't help as much as your communications do. There's always a big slump when I wait a long time for a letter. Now, I'm anxiously awaiting your letter following the signing of the armistice. I suppose you had a big celebration in Nantes. Wouldn't you have liked to have been in Paris? I would. But as it was, we had only our camp celebration and that meant almost confinement to camp because there was so much drunkeness outside.

And now everything has to be finished up and polished up before we are free again. Won't it be a grand and glorious feeling to be out of the army and our own bosses? Of course, I'm not troubled specially by bosses but it will be fun to be away from regulations other than those of my own manufacture.

Are you still getting trainloads of wounded? We haven't had a convoy now for forty eight hours and maybe more. . . .

Emma Matzen has just had a letter from her sister. She is at Ft. Dodge and they are having much trouble with flu there. Also one of my patients passed thru Fort Dodge. I felt a sort of kinship with him. Funny! Funny to be all so wrapped up in the doings of one other person!! I must be dippy.

Later

I've just come from a Jazz concert at the Y.M. Hut. The music was good—but how I do hate old jokes, don't you? We've some really good musicians among the boys and they have an excellent jazz orchestra.

Are you hearing regularly from home? I find that helps a lot in keeping up my morale. But that doesn't help as much as your communications do. There's always a big slump when I wait a long time for a letter. Now, I'm anxiously awaiting your letter following the signing of the armistice. I suppose you had a big celebration in Nantes. Wouldn't you have liked to have been in Paris? I would. But as it was, we had only our camp celebration and that meant almost confinement to camp because there was so much drunkeness outside.

M'deah, won't you be happy when censorship is done away with? It remains a trial for me even after eighteen months of its beneficence. Theoretically it's all right but when I see the officers censoring it in the office something just rises up inside me against any such men breaking into my private correspondence. It won't be very long any way. . . .

Letter # 121

Laura to Erle
France
Nov. 18, 1918.

Dear Buddy: —

. . . You know, m'deah, I know so little about your camp—what it contains. In fact the only thing I know about it is that it is in the grounds of an old chateau and that you are in it.

Tonight we heard again that Besley is trying by his wire pulling to get us taken back to the States and into a hospital there as a whole unit. It makes us all swear. Why can't that cur leave us alone now that he is away from here? And I can't quite imagine being in a hospital there with our present outfit.

I think I'll put in for a transfer to A.E.F. preferably to Hospital 11. I wonder how that would work out. But someway I have a hunch that things will work out all right anyway. I don't see any reason for them not doing so in the course of a few months.

Oh won't it be lovely to be out of the army? I'm finished with all things military for the rest of my life. . . .

Tuesday [***Nov.19***]

It is very hard to keep right on going just as busy as ever and with no visible difference in anything since the armistice. I had imagined that within a week we would notice some difference but if anything we are busier. The last evac. we had was the night the armistice was signed. And we keep on getting convoys. We've had to open up some of the wards that were closed down for the winter. They are wards without heating or lighting facilities and unfit for convalescent men of any sort. But we must make room for the stretcher cases that come in. As a matter of fact there should be no such wards in connection with any hospital any place. I hope you have nothing like them in the American hospitals.

Did I tell you that Frances will probably be sent home as she has to have six months rest. She was over last night and needless to say is all in a flutter. She is terribly nervous but of course she always has been.

You should hear her tales of the discipline in the hospital. It is awfully funny because Frances never has done anything much only just what she has pleased to do. And the discipline is getting a late start with her. But she is discouraged about going back. . . .

I see in the Daily Mail that American censorship is to be lifted. I'm glad of that but it won't help us much as long as we are with the B.E.F. . . .

Private Baker's memoir:

November 17

Got my bunk and clothes together and have settled down. Took a hike with a buddy to Paris-Plage during the afternoon and visited with my friends, the DeSchryvere's who run a store. They thought I had been killed as it had been five months since they had seen me...

I have not been paid for three months but I had no chance to spend anything at the front anyhow.

November 18

Got paid to date. Lose my Surgical Assistant appointment and now am a Medical Assistant. There are no more doings in the Operating Theatre [*British spelling*]. Wounded are left here and will get the best of care as long as they need it now. There are many terrible cases lying near death. One poor fellow has four or five tubes draining an open wound in his lung and chest which has become infected.

Some wonderful plastic surgery is being done here and at #4 across the street from us. New bones are being put in noses and some have new aluminum jaws.

November 19

We do not have much to keep us busy. We only have to keep the camp in order and spend eight hours in the ward with the wounded.

Letter # 122

Laura to Erle
Picardy
Nov. 22, 1918.

Dearest Buddy: —

Today I received your letter of Oct. 31st and again I'm puzzled by your reference to the "stinger in the letter of Oct. 24th". What in the world did I say that hurt so? Whatever it was must have had an ambiguous interpretation and you chose the wrong one. Please read it over and try the other one and see if it doesn't make you feel better. That's the only solution I can offer because I haven't the faintest idea what I said but I'm sure that I didn't intentionally send a "stinger."

A day or so [ago] I had a letter from Miss Powell that I'm enclosing. Also I'm sending Lt. Murray's last letter. The latter isn't specially interesting but it is friendly. . . .

From Lt. Robert Murray

Canadian Convalescent Hospital (Officers)

Matloch Bath.

Derbyshire

3/11/18

[*November 3 in English style*]

Dear Laura,

I wrote you & Frances from hospital in London, but it seems you didn't get my letters, as I've had no acknowledgement.

I did not have an operation, so the 'foreign matter' remains—the result of mixing with those Germans, etc.

After I was a few days in hospital, I was allowed out from 11 A.M. to 10:15 P.M. every day. It seems a remarkable thing for an English hospital, but I think it was on account of the rations being so scanty—the food was wretched.

Just twenty days after being wounded I left the hospital convalescent, & am now at a hospital in the Midlands for Canadian convt Officers.

It is one of the swell places of England & really it is glorious. Lots of woods & hills, & a beautiful stream. If it were just Summer so I could go fishing.

I expect to be here only a short time; I notice that as soon as a fellow can get around at all, he is sent off marked General Service.

Isn't the news fine these days? It is quite on the cards that I won't have to come to France again—I'm not going to volunteer this time—at least that's how I feel just now, but I believe a depot in England is a hotbed of misery, so anything is possible when I report for duty there.

How are you & Frances getting on? I hope Frances isn't giving way too much to her addiction—poor dear!

I expect to be on leave shortly. Please write to this address—c/o Mrs. Alec Russell, 9 Douglas Place, Galashiels, Scotland, my sister's, & let me know when you are likely to be on leave as I intend to send you a cake of real shortbread from Edinburgh, if obtainable; & I want you to be there to receive it.

Please give my regards to Frances. If there is anything I can do for either of you here, don't hesitate to mention it. I asked Frances in my last letter if she really wanted a faked letter. If she does—tell me in your letter, which I will expect from my sister in a few days, if I'm not there to receive it.

There are lots of Americans in London. I'm afraid they will be getting a poor impression of the city; it is rather a wretched place these days of rationing, & it is getting to be the foggy season. I don't know if you have fogs in the cities of America, but no doubt you've heard of the London ones. I always imagined they were exaggerated until I experienced one last fall. Cost me an effort to push my way through it.

I hope you are less busy now that you are on days again. I thought you were looking just a little bit harrassed last time I saw you.

Best of good wishes to you.

Sincerely yours. Bob.

Private Baker's memoir:

November 26

Flu is very bad. Hundreds are dying all around us. We act as embalmers. I have fine health so far.

November 27

"Unit" dance held in evening. I took Miss Van Ark of the staff. A lot of boys brought girls from Hospital #22, also Americans. We had a fine time. Only three officers were present.

Hospital is full of flu. Every bed is taken and there's no relief in sight. Men have low resistance after four years of war and catch the flu easily. We have lost one officer, one man and one nurse from it already.

Letter # 123

Laura to Erle
Camiers 1-12-18
[December 1]

Dear Buddy Boy: —

I'm so very, very sorry to hear of your loss [***death of Erle's father***]. And it must be so very hard for you and your folks to have you so far away at the time.

I had had no letter from you for so very long—Nov. 5th the last one—that when Bertha's letter of Nov. 6th came enclosing a clipping from the local paper telling of your father I was much ashamed that I hadn't written oftener. But you see I write awhile and then I sort of lay back and wait awhile for an answer. And I've been waiting all week. Finally last night arrived yours of Nov. 12th and Nov. 21-24 and also the one full of clippings.

I am much relieved to hear that your ear is all right again, that you haven't had the flu and that not more than one tenth of your unit had the flu! We didn't have nearly that high a percent here.

There is a nurse here from #34 hospital and she is going back tomorrow night so I am sending this letter by her to forward to you. She is a Miss Robinson and I knew her while in training school. She tells me Nantes is well worth seeing and offers to entertain me while there. But I think I shall not accept the offer. I prefer a hotel and freedom from obligations if possible. . . .

11 P.M.

Miss Robinson has been in here all evening talking up Nantes to me and I'm getting really all excited over it. Only I have no clothes and that is a tragedy when one wants to go avisiting. Our suits ordered in the Spring have not yet materialized because no one in charge cares a darn about anything but themselves and our hats have not come and altogether it is a cold, cruel world. But I'm coming anyway regardless of clothes or rather lack of them. . . .

When I get out of the army I hope I'll not land someplace that will expect such punctuality. But I'm sure I don't know where that would be. . . .

You may expect me in Nantes sometime around Christmas probably about the 26th or 27th. If we leave here on the 20th I don't see how I can make it back there before the 26th and I just don't dare to come there first. . . .

The Sweethearts Celebrate

Letter # 124

Laura to Erle
"Rec'd at Nantes Dec. 14, 1918,"—two days after receiving a letter dated five days later.
Picardy, Dec.3, 1918.

Dearest Sweetheart: —

. . . I'm going to pretend that Nice is not on the map and spend my whole leave in your city. . . .

Two of the units that came when we did—also attached to B.E.F.—have received movement orders. We have also heard (by rumor) that we will sail from Brest.

My roommate thinks I'll always regret not seeing Nice, Emmy Matzen says three days in Nantes will be all I'll want and so they all advise except one who is also in love and seems to understand. And anyway, dear Heart, how could I enjoy the Mediterranian and its blueness when it is you I want to see. I'd feel awfully cheated. And if I try to go both places, I'd simply have to travel all the time and travel in France is a nightmare.

If I get my leave on the 20th you can expect me on the 22nd. I'll stay a day in Paris to purchase a hat and will take a night train out if I can so as to arrive in the morning. I'll go to the Central Hotel (on advice of Miss Robinson).

There is such a delay of mail that I'll write this again to you in another day or two. Perhaps one or the other will get thru. I sent a note by Miss Robinson to mail in Nantes and perhaps you will get it soon. . . .

Do you realize . . . . that we may have a Christmas together in France? I hope your work continues slack so we can have some time together. Now, my disappointment will be most horribly keen if we all are turned out of here within a week or so. But being in the army, I realize that individual desires have nothing whatever to do with the movements of units. Please pray hard that we won't go for awhile. I don't want to go back before spring now because you won't be there and things will be so blank.

I understand that our letters are not being censored just now so I'm writing this as if I thot so. But I'll send it thru the Base and another thru ordinary channels very soon and surely you'll get one of them. . . .

Nothing else of importance has occurred. Frances has been boarded and is to stay on with us. One of the others was boarded and is to leave Thursday. That doesn't sound as if we were all going within two weeks, does it? . . .

Letter # 125

Laura to Erle
Dec. 5, 1918.

Dear Buddy: —

If you receive my last two letters you'll know I haven't an ounce of decision and when you get this you'll be irrevocably convinced of it. I bethought me of a calendar and discovered that I would have no Sunday with you if I left here on the 20th and also that the Saturday before Xmas would be a very bad time to shop in Paris. So I asked for the 19th instead. That will make me reach Paris that evening and will allow the 20th to purchase a hat and get on the trail of the long lost suit. And then I'll take the first available train. I can get no information about trains out of Paris. But I'll have to have at least the morning to shop and maybe part of the afternoon all depending upon how I find things. Then I'll hunt a train. Not knowing how trains run nor how long the journey is I can give no definite information. I hope you will be able to get some off duty time the first day I'm there else I'll feel rather beset but, of course, we'll have to make the best of circumstances. I'll go to the Central Hotel. . . .

Today I went down to the Chief Nurse's Office and asked what the prospects were for movement orders for leave. She said she thot we would get them but she couldn't guarantee that we would. And we may not get them on exactly the desired date.

Now if we would get orders to break up the hospital and return to the States immediately, my heart would just naturally crack with disappointment.

Yesterday I went to B___ shopping. It was an awful trip. The trains are so polky . We were an hour and half going each way and I'm sure it isn't more than fifteen miles. But I do not go often so should not complain.

Tonight I received my Xmas box from home but I haven't opened it. I think I'll save it a while and see if I get my leave. If I don't get my leave I'll celebrate Xmas by opening the package. Has yours come yet?

Yesterday going in on the train I talked with an English Y.M. woman who seemed to be very familiar with France. She said Brittainy is charming altho the winter climate is beastly. She thot Mont St. Michael the very prettiest spot in the whole country. I've looked it up on the map and find it is quite a way from Nantes. . . .

Tonight we got the Indianapolis and Chicago papers telling of the perfectly wild peace celebrations. . . .

Today I went down to the Chief Nurse's Office and asked what the prospects were for movement orders for leave. She said she thot we would get them but she couldn't guarantee that we would. And we may not get them on exactly the desired date

Now if we would get orders to break up the hospital and return to the States immediately, my heart would just naturally crack with disappointment.

Letter # 126

Laura to Erle
"Received at Nantes December 12, 1918"—earlier than previously mailed letter
Sunday, the 8th

Dear Man: —

How many letters have I sent you now announcing and re-announcing the fact that I'm coming to Nantes if I get a leave? . . .

I've never gone anyplace all alone before since I've been in France and occasionally I have a cold wave trip up all my enthusiasm. I can't speak French at all and I anticipate a little embarrassment but suppose I can manage to get there in one way or another.

I know a Miss Robinson at #34 and I suppose I had best write to her also and let her know when I expect to arrive. By the way that will be sometime on the 21st as I expect to be in Paris on the 20th. I leave here on the 19th—if I get my movement orders when I asked for them—reach Paris that night, shop (in reality buy a hat & look up a long lost suit) and then I'll be free to start to Nantes. I do not know and can get no information about trains between Paris & Nantes so I can't say what one I'll come on but very probably you'll be working anyway. I'll go to the Central Hotel (recommended by Miss R.) and you'll find me there sometime on the 21st. That is Saturday and only thirteen days from now. I wonder if that is as exciting to you as to me? . . .

The powers that be have promised to move me away from the ward I'm in and I'll be so happy to move. I haven't been satisfied with the ward since I've come off night duty and it will be a pleasure to move. Of course for thirteen days and in reality it's only eleven till I leave here. . . .

We've been getting plenty of coal recently and the necessity of scrounging it (I refuse to call it stealing) has passed. We have a specially good heater in our room and we are very cozy.

I got my Xmas box from home this week and couldn't resist opening it. It had candy & handkerchiefs and hair pins and chewing gum, two prs. of woolen stockings and couple of small gold collar pins. Wasn't that a lot for such a small space?

I suppose the folks will be expecting me home very soon. And in fact our chief nurse said today that #5 Hospital Unit has orders to go in January. They came about two weeks before we did and were located in this camp for a while. I'm so afraid our orders will come before I get away on leave and now I would hate to go back to those United States of America without seeing you! . . .

Erle to his mother

Sergt. John Erle Davis

USA Base Hospital No. 11

Amer E F APO 767

Nantes France

Christmas 1918

[*Postmarked 12/10/1918*]

My Beloved ones —

It is a pity of course that I can't be with you in person to help welcome Santa Claus, but it is a real comfort to know that we are together in spirit. The actual meeting and visiting isn't very far away now, and you'll be hearing the mountain of things I've wanted to tell all about in letters — but just couldn't.

Naturally, everybody in the A.E.F. wants to be first to start for home — it is very entertaining to us to hear that the French fear most of us expect to remain "over seas" and take their jobs, their money and their wives and sweethearts, when in fact we wouldn't trade our country & our American birthright for all the rest of the earth.

Rumor says that our outfit will leave for the States early in January and that by early February we'll be out of the army — visiting around, lying frightfully about our marvelous experiences, and presently, hunting jobs and settling down to the old grind.

At present the most popular man in camp is a patient who tells fortunes with cards — and he's very good. The nurses want to know if they'll be married soon and if their true loves have been really true. And the men are all afire to know if, in their absence, the girls back home have gone and married "some other guy." Fortunately I have no worries of that sort to plague me.

Yesterday's mail brought a number of Christmas boxes. To me — a beautiful handkerchief in a letter from Carita, and cards from three of four other friends.

I didn't send home a certificate for a Christmas package. Don't ask me why — I don't know! If I get a letter from you all I shall be satisfied. I haven't sent any presents. You'll have to wait until I come. Some of the boys have sent stuff homeward and it has been lost in transport in many instances. I prefer to take no chances.

I haven't kept in touch with anybody in America except you and the Rees household. At first I sent cards to several friends, but the pile of letters received have gone unanswered. There's so little to write that's interesting which isn't forbidden by the censor. And trying to say something worth reading, while considering censorship every second, simply suffocates me mentally and spiritually.

Some of our fellows are on furlough now, visiting some of the famous places in France. I hope I shall have a similar opportunity soon. But my job is such that it may be impossible for me to get away for more than a day at a time. (I am the non-com. In charge of the Registrar's office, said to be the most important office in the works. Previously, I was in charge of the evacuation of all patients from hospital.) Failing in my desire to go on a leave, I hope to have Laura run down to Nantes for a couple of days. Several weeks ago I invited her to have a look at our fair city — it is a famous and interesting and picturesque place — and she said she'd come if she got the chance. We

I haven't kept in touch with anybody in America except you and the Rees household. At first I sent cards to several friends, but the pile of letters received have gone unanswered. There's so little to write that's interesting which isn't forbidden by the censor. And trying to say something worth reading, while considering censorship every second, simply suffocates me mentally and spiritually.

have been steadily in communications since she left America, though I've not had a line from her in the last two weeks.

About my future activities I'm rather cloudy. I aim to go to Washington and line up with one of the news bureaus — perhaps the Philadelphia Public Ledger — and within a year or so pry my way into a place as Washington correspondent. I'm not keen to remain in Chicago. Opportunity there for me is not big enough — Chicago would laugh at that, but I feel it's true so far as I'm concerned. Within a few days I will write to friends in Chicago and Washington and put them on the trail for me. You see, I want to have a place ready for my presence when I get home.

Beyond anything else — I trust your Christmas is happy. If love without limit means anything, you will be joyful. . . .

Note that Erle does not mention in this letter that he has invited Laura down for Christmas from her hospital hundreds of miles away on the French coast of the English Channel.

Laura and Erle became formally engaged after a Christmas Eve midnight mass at the cathedral in Nantes. But oddly there is no letter from her to him signifying the event. And because of religious differences—he was Catholic, she Methodist—both were reluctant to inform their parents.

Private Baker's memoir:

December 24

There was a "unit" party held at which we all enjoyed real American music. Our wonderful jazz orchestra was at its best and one patient, a cabaret man from Chicago, gave us a lot of real jazz songs.

Our Commanding Officer, Major Nusbaum, announced that we would be leaving France within thirty days. There was wild applause for we sure appreciated "noise" like that.

Money from Chicago friends supplied the whole unit with a real Christmas dinner.

December 25

Got up at 5:00 a.m. and about fifty or sixty of us sang carols all over the hospital. All of the patients sure did love it.

There was a fine spirit in the hospital during the whole day. Miss Judy's ward was a regular Christmas menagerie. The boys from my wards, among whom were a couple of professional vaudeville actors, gave a trick show and I took them all through the hospital. A six round boxing match took place in Capt. Nadler's ward between two Tommies. The Americans in all the wards had a good time. There was a convoy of 150 Yanks on Christmas morning from other hospitals and they all got a big chicken dinner.

Letter # 127

Laura to Erle
Tuesday evening [Jan. 1, 1919]

Dearest Buddy

I'm sending this just to let you know that I've arrived safely. I missed my train yesterday morning so padded about Paris all day looking for something I wanted to purchase. At the end of the day my total purchases amounted to 2 lbs of coffee (um-m!), one cup of tea, one dish of ice cream.

I met the other girls at the Y.W. so came on home with them and got in at seven this morning. They aren't busy so gave us the day to recuperate and I've put it in talking and going thru my trunk. . . .

Rumors are rampant but the general supposition is that we will go within a month. The patients are few and none coming in and it looks as if we were going to evacuate shortly. One of the hospitals in the camp is clearing out and leaves here on the 8th. One rumor says we won't go before the 20th but I don't see how we could go before that anyway. . . .

Letter # 128

Laura to Erle
Jan. 4, 1919.

Dear Mr. Davis: —

It has rained almost continuously since I arrived in camp. Tonight there is a patter, patter on our roof. I wonder if it is so in Brittany or are you all flooded out? If it has rained as much there as here since I left I'm sure that you have moved and left no address.

Have you missed me since last Sunday? No? I can imagine that you needed a good long rest after my departure. Never mind, dear heart, I shan't come again to disturb your labors.

We've been having a round of dances and entertainments here and of course I've been to some of them. New Year's Day we had an At Home in our mess for the whole unit. The cakes and coffee and cocoa were excellent—also the sandwiches. Last night I went to a dinner at Etaples at Tipperary Hut, the largest [***hut?***], with a colonel—who said "Really" to each of my remarks.

Today I've been loafing again and going thru my trunk and looking at my pictures. . . .

I sorted out a bunch of stuff to send home by mail. I understand that we are to be quite limited as to baggage so I'll have to get my clothes home by parcel post.

The air simply buzzes with rumors today. I rather think that at last a definite date of leaving has been received but I don't know. I feel sure we won't be here longer than a month and I'm getting very anxious to get started. I'm taking no interest at all in the work of the place.

We hear that a British unit is coming to relieve us and I don't care how soon they come. . . .

Censorship now has officially relaxed here. It was welcome news even tho I know that censorship was very nominal here since the armistice was signed. . . .

Sunday Evening [***Jan. 6***]

Now, I can scarcely wait to get started home. I keep thinking about reaching Chicago and North Vernon and what I am going to do when I get there and afterwards and about our plans to meet and everything. And everything just must turn out all to the good "Papes" [***reference to their religious faith differences***] and all.

I anticipate considerable sport on that subject. But you know, honey, they all love me and I'm the youngest child and they'll mean it all for my own good. . . .

The men had to hand in their barracks boxes today and most everything else. They had been told they would be allowed forty lbs. in their barracks bags, the sergeants 100 lbs and the nurses 150 lbs. . . .

Typewritten letter from Erle to his mother and sisters in Indianapolis

Brittany

Sunday

January 5, 1919

Dear Girls — all three of you!

The Christmas box came New Year's day, and Hetty's letter of December 10 came day before yesterday. For both I rise to utter thanks. Since the middle of November, when I heard of Dad's death, I have had almost no mail. Of course, we Americans in Europe certainly had no reason to kick if the government filled valuable cargo space with munitions and supplies needed in the zone of conflict instead of letters from the home folks, while the war lasted. But some of us are inclined to be impatient with the postal-express system since the signing of the armistice. The mail service hasn't improved.

But, ho hum! Who cares? We're starting for the United States within a few days. If those who remain behind don't get their letters promptly, or at all, let them kick. Ours, you see, has become a policy of complete indifference with regard to the other fellow's joys and sorrows.

About our going home? Well, in point of foreign service it seems that. we are considered an old organization. The government is sending the older outfits home and relieving them here with units but recently arrived from the States. In the large hospital center near Nantes there are three units that have been on the job at least six months. Ours is one of them. The three, so I'm told, will leave France together. The relief unit. for us will arrive this afternoon. They - the staff and personnel - arrived in France November 1. Lords knows when they'll get back to see the Goddess of Liberty.

The day the armistice was signed our hospital was comfortably filled. That night - while the whole of France was whooping things up in celebration, we received two trains - about 600 patients. About

(over - stationery is a luxury. I must be frugal)

half the cases were very bad; five or six died on the litters in removal from train to hospital. They had come directly down from the Argonne front. We have now more than 800 patients—but that's a very small number for us.

Did you hear me say we shall be glad to sail toward the Western World? Quite true. There is more than one reason for the emotion. Naturally we're dying to see our loved ones. Then most of us know that we have to do some tall job-rustling. We read in the papers that the people of Dear Old America are going to see to it that all the boys returned from overseas are recipients of jobs better than those they left when the tom-toms of war sounded from the hilltops. But we're cynical. Yes. Long experience with the human race leads to the belief that after the first several boatloads reach home ports, after the first several parades have been made, after the tears of the first several public sobs over the dear heroes have been dried, the home folks very soon will become accustomed to the sight of returned fighters, and accustomed once, soon afterward will become bored. And so, selfishly, want to be among the early ones returned. Our eyes are on the meal tickets.

May I be permitted to say none of us will grieve when France is no longer visible to the naked eye? The reason? Here it is:....National color of France—Rain; national coat of arms—Rain; national sport—Rain; chief industry—Rain; principal product—Rain; national college yell—Rain; national curse—Rain; national joy—Rain; national anthem—Rain. And so forth. And so on. When we came to this spot, we were situated about a mile from the river. Now we're about fifty feet from the stream—and we haven't done the moving. Sociable river. I think there isn't a roof on the place that doesn't leak. Fact is, they aren't roofs, they're colanders. We've worn hip boots almost steadily, until our feet feel as though they had been boiled over a slow fire.

The Christmas season for me was very pleasant. I felt myself one of the most fortunate persons in the A. E. F. Laura came down from her station, 300 miles north, the evening of the 2lst of December and remained until the night of the 29th. The powers-that-be gave me freedom from duty every afternoon and evening during that period, and she and I made the most of our time together. Nantes is a historic ville, and we

May I be permitted to say none of us will grieve when France is no longer visible to the naked eye? The reason? Here it is:.... National color of France—Rain; national coat of arms—Rain; national sport—Rain; chief industry—Rain; principal product—Rain; national college yell—Rain; national curse—Rain; national joy—Rain; national anthem—Rain. And so forth. And so on.

wandered around, seeing places and regions I had not had opportunity of visiting previously. We peeped into ancient churches; looked down upon the populace from crumbling towers; groped about in dripping dungeons; explored strange passages; dined in curious restaurants; ate mysterious food; drank wicked wine; held hands in holy spots; met extraordinary people; talked incessantly, for we hadn't been together in more than eighteen months.

It took a fair measure of persuasion to get the lady down here, because in the old conventional world—now ages passed—demoiselles customarily didn't travel long distances to visit young men unrelated to them by ties stronger than mellow friendship. But in these parts we care not a damn for conventions so long as our spirits are functioning readily. Nevertheless, I charge you keep the information to yourselves, for the world at home is full of fools and wagging tongues; besides her people, she thinks, would look upon the sojourn unkindly.

There is a quaint place over on the river bank, wholly picturesque, where M'sieu and Madame Raginaud serve the most savory food in France—though all the cooking is done over a dinky flame of twigs in a fireplace built about the time of' Looie the Citron. Here we dined oftener than elsewhere. Here, on Christmas night, Hibbard, one of' my cronies, (who claims deah old Hahvud as his alma mater and who never worked a stroke in all his days because he inherited his) gave a dinner to a party of eight in Laura's honor. It was a charming affair. Christmas Eve, he and Deaderick (another pal) and Laura and I dined at Raginaud's and then went downtown to midnight mass at the Cathedral. The great old Gothic structure, with its massive columns, deep groined roof and huge paintings, dimly lighted all, was an inspiration. The service was not noteworthy. Of music there was only a dismal chant by a few toothless monks, except for magnificent bursts from the grand organ now and then. We didn't stay until the finish. For Laura and me the service was a sort of memorial - three years before, we attended a truly beautiful midnight mass at the Paulist church in Chicago, where music, ceremony and decorations really satisfied the soul. More than 500 American soldiers attended the mass at the Cathedral. As the doors didn't open until 11:30, the crowd waited and walked to keep warm - in the plaza before the church. The fine behavior of the Yanks was in contrast with that of the "Frogs", many of whom were well corned.

Christmas afternoon, Laura and I started out for an exploration of a quarter of the city I knew nothing about whatever. Presently we discovered two be-fezzed French colonial wounded soldiers following us. They trailed for more than a mile. When we stopped they stopped. When we walked fast they speeded up. All the while they quite frankly wondered at and admired her dead white skin, her wavy, coppery hair and her three gold service chevrons. At length they asked in French if she were American—an "Americaine infermiere" (military nurse)—and if all American ladies had such white skin, and such "tres bien" hair. Too, they wanted to know if the chevrons indicated that she had been wounded, or if she were an officer. I asked it they were Algerians.

Ah, no; said they, with an air. They were Tunisians - good comrades of the Americans who were good soldiers with very great courage. Finally, we shook hands all around, and they went their way happy—and possessed of wonderful news to tell their folks back home in Africa, about the queer men and women from the other side of the earth whom they had as comrades in arms in France. It impressed us really as a fine and simple bit of good fellowship, a real touch of the old human stuff.

The evening of her departure, my guest and I went out to the royalist quarter, a decidedly conservative, substantial part of town. It was great business poking around in the misty rain among old houses and narrow streets and quaint little chapels connected with monasteries or convents. We found one small church on a hill top overlooking the harbor. The outside walls were in a crumbling state. The interior was in very good condition—Gothic, ivory stone, plain leaded windows, gray tile floor. The altar was plain, just a sort of a table. But spread across behind it, like the ribs of a lady's open fan, were the flags of all the Allies. At one end was the tri-color of France; at the other, Old Glory, but much larger than any of the other emblems. That told us definitely how the French feel about it among themselves. I'm fairly sure this arrangement wasn't made to kid visiting Yanks, for I doubt if any other Americans had been in the place—it was so far out of the path of visitors. Then for dinner we went to a restaurant in that quarter, an unusual place, even for Nantes. How we found it, I don't know. To get to the dining room we went down a long, narrow hall and through the kitchen. The table linen was nice and the dishes were civilized—frequently here one's coffee is served in a cup without a saucer, and one gets a two pronged fork, and a knife pointed and quite sharp, more like a paring knife. Also there was a large fireplace, the biggest I've seen here, and in it a couple of logs burned. The place was inhabited largely by French officers and their wives or sweethearts. No other Yanks were present, probably none had been there before us, and the attention given us was flattering indeed.

While I think of it, let me suggest that you look up the September and October numbers of The Yale Review. In one or them you will find an article "Reims Cathedral" by Ralph Adams Cram, the noted architect. It is one of the most stimulating articles I've read in almost a year. Similarly it impressed Hib, Dead and Laura. Generally speaking, I find the Yale Review worth reading. The stuff is well written, whether or not you agree with its substance.

A number of' our men have had leaves recently and have seen many interesting parts of France. Some returned this evening who had been up Chateau Thierry way, up to Soissons and around St. Mihiel. I should like to have had a leave, of course, but I heard, via the "pipe line" we should be departing soon after the new year and knew there'd be a vast amount of labor in the office which would prevent my getting away. So. I took the afternoons and evenings to visit with Laura—and I'm damn glad I did so!

I certainly guessed right about the work here. We've so much on hand we have a day and a night shift doing it, in order to get all records in shape for our timely departure. Just when we'll go I haven't an idea, but rumor says the get-away will be made not later than two weeks from today. We hope we'll go directly to the States, and quite as straight to Chicago from the port of debarkation—then demobilization and home to the folks. I wish could tell you to look for me about the 10th of February. As I see it now, I'll spend a few days in Chicago, upon return, looking over the ground and renewing old professional connections. Then I'll come to you for a couple of weeks; have a week or ten days in Louisville, and step out then to find me a job somewhere.

With or without justification, I'm not very sanguine about landing a position as good or better than that I left. I rather feel that employers, knowing that returned soldiers must have work, will be unwilling to pay more than $30 or $40 or $50 a week for the service they had to pay $50 or $60 or $70 a week for before we got into the war. My feeling is shared by most of the men I've talked with, both detachment men and patients. The boys all seem to think they'll get a lot of hurrah and bunk when they get back to the States, but darned little substantial encouragement. Looking ahead, I've sent a few letters to friends who might keep an eye peeled for me. I really hope to get down to Washington on the editorial bureau staff of the Philadelphia Public Ledger. Failing in that, I will try for something worth while in the publicity line. I will not return to the staff of the Western Newspaper Union, because I know I need a change of scenery. I had outgrown the place some time before I left it, and only the knowledge that I should be leaving for service in the army soon kept me from seeking other fields before I did so. The little work I did successfully down in Washington and New York gave me a ravishing taste for "live meat."

Hetty, I'm glad that you've found satisfaction in your work at Ayers' and that you work with civilized people. As for improving your labor, there's only one thing to do: read good current stuff, write, write, write, and study well written ads. The Marshall Field ads in the Chicago Tribune and the John Wanamaker ads in the New York Evening Post are excellent. Be chary of weird word combinations, that's as silly in the long run as the use of trite phrases. Don't be afraid of writing from an unusual point of view. Trained imagination is as valuable in writing good ads as in writing good fiction. Do you read "Printer's Ink"? It's good.

You have wondered about my writing long letters to other folks and not to you? I haven't done it much. This is the fourth long one, I think. The others went to Dad, to Carita, to Aunt Sadie. Laura and I have kept in touch with fair regularity. All other friends, including relatives, have received only postcards or very brief notes, or nothing. Carita deserved the one letter she got, the one you read. She has written frequently and at length. She has sent magazines, clippings and a check—the last I was sorry she sent, of course. I haven't written Miss Chandler. I simply haven't had the time and haven't been in the mood to write much to anybody. When one has finished hammering the machine all day, or has prepared reams of stuff to be typed by others, he hasn't much zeal left at night for social correspondence.

With regard to Miss Chandler: have you had her out to the house for tea or anything since Dad died? What her feelings are toward me appears to be of small consequence. Gad, she'd be hurt if she read these lines!

I'm non-commissioned officer of the day these 24 hours. That means I shall have to make an inspection of all the wards in a little while—at 10 p.m. So I'll bring this missive to a close and make ready to start forth. By the time you receive these observations, it will be too late to make reply. Just wait and tell me all about things when I reach the bosom of my family. . . .

The photos in the Christmas box made it priceless! The fruit cake was an inspiration

Letter # 129

Laura to Erle
Camiers
Jan. 7, 1919

Dear Man: —

. . . .A British staff is coming to take over the hospital and when they come we will have to move out of our quarters and into a ward till we can get transportation. You can imagine that we aren't very keen about this British Staff showing very soon and yet I'll be delighted to get the hospital off our hands. The men have been told to pack so I suppose they will be sent shortly. We have a number of Royal Army Medical Corps orderlies here and they will probably manage till the rest of their staff comes.

I've almost finished packing up my clothes. I haven't many and by the time they are all sorted out I think I can get them all into my trunk. You should see the French women and children in our camp raking thru the paper cans that stand outside our huts. They stand and wait for some of us to [***go***] out to empty a waste basket. Then there is a mad scramble to see who gets what.

The Harvard Unit, located next to us, departed this morning and as soon as they had gone the whole village of "frogs" was swarming thru their quarters looking for cast offs. Are these the fine sensibilities of the French people? Or do they really need things so badly? It is rather a question in my mind. Undoubtedly they have suffered much. I don't like beggars, do you? . . .

I've thot of a solution to the problem of how we are going to see each other when you get home. Since you won't (or shall I say can't) come to N.V. and I won't come to your house nor to Louisville. I may go to Indianapolis to visit Edna Allen for a day or so. She'll probably ask me to. If she doesn't why of course my suggestion falls thru. If you came fairly soon it will be very convenient to do so because I've just got to have some clothes and they aren't to be had in our native city.

Tonight is the last of the Machine Gun Corps concerts at the Y.M. and I'm going. I'd be glad of your escort but since you refuse I'll find me another. They are plentiful. . . .

Patient Load Declines, along with Morale

Letter # 130

Laura to Erle
Picardy
Jan. 14 1919

Dear Buddy Boy:

. . . Some of the folks here are getting three days leave to Brussels. I put in for one but have no expectation of seeing it. I shan't be disappointed if I don't. . . .

Evening

Now what do you suppose happened? Our three day leave has gone thru! We go soon, probably about Friday and I'm going to borrow a few francs to tide me over till pay day. Financially I should not go but why miss the chance? Eleven of us are going together and if there is any thing in the Safety in numbers theory we'll be perfectly safe.

It takes almost twenty four hours to reach Brussels from here but we'll have one full day there and travel back along the old line. One of the boys who has just returned has been in telling me all about it and is very enthusiastic. He says it was way ahead of both his Paris and Nice trips. . . .

Letter # 131

Laura to Erle
Camiers
17-1-19

Dear Buddy: —

Tomorrow morning at 4 o'clock I start to Brussels on a seventy-two hour leave. We will be in Lille a few hours and reach Brussels some time in the evening. We'll have one whole day there and come back by way of Amiens so as to see much of devastated country. It will be a hard trip but it is the best we can do now and we are tickled to have the opportunity to see that much.

As yet we are awaiting orders. We still have nothing definite and as far as appearances go everything goes on just as it did the day after the armistice was signed. I don't think we'll be going this month but I can't believe that they'll leave us here till spring.

I hope that you have started on your way by the time this reaches Nantes but since army movements are so polky I hardly think you will be.

I hate to think of being out of touch with you for even the time it takes to cross the ocean but it probably won't be much longer than the way letters do now. The A.E.F. mail service is certainly rotten.

Did I tell you that our letters are no longer censored— unless they are suspicious? That helps some. [***Yet the envelope containing this letter—and the rest right up to embarkation for home—was marked with a censor's rubber stamp, in this case*** "PASSED BY CENSOR NO. 1402."]

We are signing up in groups of four for one day trips up the line, probably Ypres as that is near. I suppose we'll go in the Ford car. That will be fine don't you think so? I'll bring you a German shell case if I go. I have one but I think I'll give that to my father. Men are so difficult to present gifts to and here I'm considering adding on another male member to my family group! . . .

Letter # 132

Laura to Erle

This letter, postmarked 1/22, was "Received at Nantes April 1, 1919"— nearly 2 1/2 months later
Back in Camp
Jan. 20, 1919.

Dear Man-o-Mine: —

I've had three strenuous days of travel and sightseeing—mostly travel—and here I am again. We got back last night at eleven o'clock and simply fell into bed. We didn't report on duty till noon so had a good rest.

This afternoon I wrote home all about what I saw and I find I'm all talked out on the subject so will have to tell you about it some other time.

I heard a very good opera night before last. I enjoyed it much even tho I was so desperately sleepy I could scarcely keep my eyes open. It was "The Daughter of the Regiment". The singers were good and also they acted well. . . .

Your preparations for departure seem much more serious than ours. I'm sure I don't know what we are waiting for unless it is transportation. We continue to receive convoys and run strongly as usual except lightly. None of the wards are full and most of the cases are light.

I had letters from home, also. Silas at last is able to get along without crutches. He has a brace for his knee and with aid of a cane can walk about quite well. I was so glad to hear of the improvement. I was afraid he would never be able to walk again. You see it lacked just a week of being a whole year since he was injured. . . .

The Next Evening

This morning I went to the ward that had just seventeen patients. There is nothing much to do [connecting word missing] give a few medicines and make five beds and see that the other work is done. So this afternoon I sent an Aussie down to the village for flour and to the canteen for syrup and made up a lot of batter and baked pancakes. I wish you could have seen the bunch. They acted like so many little kids they were so tickled with them. They all had all they could eat and even more than they should have eaten but they were happy over it. Honestly, it is lots of fun to do a thing of that sort when it is appreciated. And it surely was.

There's no news of our moving. I begin to think we'll be here months yet. But we'd probably know if we were going then. . . .

I have two more things I want to do here. I want to go up to the old line for a one day trip and I want to go to see Miss Simpson. Then I'll be absolutely finished with France.

The District Director Medical Service has promised that we'll all get to go up to the line and six are going every other day. And if we find we'll be leaving soon he promises to send us all in a bunch. So I expect to get to both places. . . .

It is just a week tomorrow since our last patients were transferred out and the hospital closed. We are completely packed and daily expect to go either to England or to be shifted to another camp.

Letter # 133

Laura to Erle
This letter was addressed to "1/c Sgt. John E. Davis," indicating he had received another promotion.
Camiers
Feb. 7, 1919. [Thursday]

Dear Buddy: —

The much looked for letter arrived today—dated Ground Hog Day—and it didn't tell me you had departed for the States. I am much disappointed because you haven't gone. The interval between letters was so long that I was sure your letters had gone astray and that I would hear no more till we were both back on our native soil.

As a matter of fact we, too, are marking time. It is just a week tomorrow since our last patients were transferred out and the hospital closed. We are completely packed and daily expect to go either to England or to be shifted to another camp. Yesterday an order came for fifteen more nurses to go to Coblenz, Germany. Twelve eligibles volunteered and three more were detailed and they departed this noon. My roommate, Miss Cleveland, and Miss Pawlisch were among the volunteers so I am a widow lady tonight. Frances Hampton and Bess Gambier also volunteered but were turned down.

I feel the least bit guilty not to volunteer when I am such a husky. And if it were for six months only I'd like to go and see the Rhine country. But being much in love with a long, lank, first-class sergeant I think it best not to tie myself up indefinitely with the Army of Occupation. I am too anxious to get our little home started. Time flies so and we've waited so long for each other. And what is the Rhine country to me if I can have you?????

If I should be detailed to go I'll make a struggle to get away in the spring. Tomorrow morning we are to have our baggage tagged and according to the latest rumor we depart for home on Monday [Feb 11]. But please don't take such rumors seriously. When you get a note saying, "I am sailing within a few minutes", then change my address to North Vernon, Ind. . . .

If you should go home first and don't know (and how would you possibly know) when I am coming will you send a note when you land to this address and also one to North Vernon? I'll do likewise sending one to Unit 11 and one to Indianapolis. That way we won't be out of touch so long, n'est pas ***[n'est ce pas:*** is it not so*]*? . . .

Flag being lowered, February 1919.

Letter # 134

Laura to Erle
Camiers
Feb. 11, 1919. [Monday]

Dear Man: —

It has been bitterly cold for several days so I have simply dug in and waited for milder weather. I've read a little, sewed a little and crochetted a little. By the latter you may judge the depth of my boredom. We've been a little short of coal but by means of scrounging I've had enough to keep nice and comfortable.

After Miss C. went to Germany Frances Hampton decided to move in here with me. I'm not at all enthusiastic but I hope it won't be a great while till we go. Tonight she has gone to a dance and has left me in peace. I'm sitting in front of my nice little fire and have the lamp shade tilted just enough to give a light. . . .

We've had much snow and the hills are very beautiful.

Wednesday [*13th*]

Today I had a scrumptious walk. Emmy and I went up to the club to tea and then up the hill to the tip. We couldn't see far out because of the mist in the distance. The sunlight was all reddish out on the water and the two light houses at Paris-Plage were barely visible. Then we walked down the hill toward Etaples and quite a way down the Etaples Road. One effect was especially good—the yellow-brown-gold (if you know what that is) sky thru the dark green of the pine woods.

Yesterday there was purplish mist over the hills but that was gone today. I wish you could see our clouds and atmosphere here. That is about all I can brag about for this district.

Tomorrow we are giving a fancy dress party in our mess—ladies only—and we are all busy planning our costumes. I have yet to make a very full short skirted affair and Frances is going to rig out in her blue silk kimona.

I saw the item in the paper about the resolution to give the soldiers of Ill. six months pay. It is certainly the decent thing to do. I hope it passes.

I've been wondering a lot about how we'll find things when we get back. As far as I'm concerned I'm indifferent. My profession has lost out in my interest and I'm not caring much about it. But for men it is an important matter and I know just how anxious you are to reestablish yourself professionally. I can't think you will have any difficulty doing so but I don't know conditions in the U.S.

I wonder if I'll ever be able to be of any assistance to you in your work?? Yes, I know very well that I know nothing about it but that doesn't permit [***prohibit?***] me from wanting to help and wondering if I can.

I was so disappointed this evening not to get a note from you. I had set my heart upon getting one but nothing appeared at all.

Did I tell you that I have some souvenirs for Fred and Gene? I'll send them on some day when we get back to the States and I have their addresses. They are just alike so no jealousy can arise! I want you to see them first and if they are all right before I send them.

I'm so sorry to hear of Fred's transfer. But cheer up, my dear. A transfer isn't such a horrible tragedy in a war. Who is the fourth in your quartet? Carl? I don't remember the other name but you know who I mean.

I've forgotten to ask you about the welfare of the pawry [*?*] Miss Aldine. I hope she recovered from her cold and blues. The diseases sort of match up, don't they?

There has been a rumor about that we may have to open up the hospital again because of a return of the Flu epidemic. I hope we don't. I'm all finished up mentally and don't want to ever see a ward again. Lazy? Yes and bored.

Do you realize that it is twenty one months since I left Chicago? You know that is quite a while for me to be away from home. And it is almost two since I saw you. That was surely a bold thing to do to come a-visiting you. I wonder how I ever had the nerve.

If we go back thru England I want to get some table linen there. They are supposed to have very fine linens, aren't they? If we don't go that way of course I won't get it but I'd like to very much. I'll have to make myself satisfied with some of Marshall Field's or even with Ayre's at Indianapolis. . . .

This letter, postmarked Army Post Office S.18 13 FE19, was noted by Erle, "Rec'd at Nantes March 8-1919"—more than three weeks to cross 300 miles of France. That's longer than it sometimes took letters to cross half the U.S. and the Atlantic between Laura and Erle. After war's end, there evidently was a system breakdown in the mail service for American forces in France.

Letter # 135

Laura to Erle
Camiers
Feb.14, 1919.

Dearest Buddy: —

. . . But to get down to business, I didn't get a letter from you and I'm greatly distressed because we have our movement orders at last. They came today.

We—the nurses only—leave here on the 16th and go to Vannes. That isn't far from Nantes and I'd love to stop off and say Hello. But not having heard from you for a very long time—a week to be exact and the letter was dated the 2nd—I think you may have sailed for parts unknown or else have stuck in the mud.

Yesterday all the casuals who had come to the unit as reinforcements were sent to reinforcement camps to finish their service in France with some other outfit. It distresses me much. I'll not feel at all settled till I know you are safe in the U.S.A. . . .

Private Baker's memoir:

February 14

I secured an all-day pass to St. Andre and I left at 4:12 a.m. I went east through St. Omer, Hazebrouck and Armentieres and arrived at St. Andre at 10:30 a.m. My purpose was to visit the Bigots and I got there before my letter had arrived. They did not expect me so soon but I had a warm greeting just the same from Mother Bigot, Louise, Jeanne and the two dogs. These people, as stated previously, are the French people I met during the war while we were advancing through their territory and who had just been released by the Germans after four and a half years. They had suffered from all the terrible bombing raids of the English and from the awful concussion of the great long range guns of the Germans not far away. They lived in Paris during the last war [*Franco-Prussian, 1870*] so Mother Bigot remembers the Huns well. They are of the better class of French. Mother Bigot's son, who was a doctor, died just before the war. When I came there first, they took me in and treated me just wonderfully. They gave me the very best food possible and food was scarce as gold.

The mother just boils in anger at the name of the Germans. She had buried all her silverware and bronze fireplace fixtures, etc., before the Huns had captured the town. I was the first guest to use them after they were dug up. They will have to be cleaned as there are deep stains on them caused by the moisture in the ground where they had been buried for five years.

After a regretful farewell, for I knew it would be my last visit there as we were awaiting orders any minute to pack up for home, I left them at 5:40 p.m. I got back at 12:30 a.m. and call it one full day but one which I shall long remember.

The train runs through miles of the worst devastated area in France and it has to run slowly on account of the many shell holes and the fact that the track is not very permanent yet. There were hundreds of Hun prisoners cleaning up this area while I was there. Armentieres, as well as a lot of the ruined towns, are full of civilians now.

Nurses leaving for home.

Mother Bigot remembers the Huns well… Mother Bigot's son, who was a doctor, died just before the war. When I came there first, they took me in and treated me just wonderfully. They gave me the very best food possible and food was scarce as gold.

The mother just boils in anger at the name of the Germans. She had buried all her silverware and bronze fireplace fixtures, etc., before the Huns had captured the town. I was the first guest to use them after they were dug up. They will have to be cleaned as there are deep stains on them caused by the moisture in the ground where they had been buried for five years.

Letter # 136

Laura to Erle
"Rec'd at Nantes March 17-1919"—over a month in transit
Feb. 15 1919

Dear Honey Boy: —

. . . We are all packed up and leave tomorrow for Vannes which isn't far from Nantes. We suppose it is to await sailing but do not know. Only the nurses are going. Neither officers nor men have received orders.

I so hope you'll go across soon. I'm going to be almost unhappy till you are back in the U.S.A. . . .

I'll write you from Vannes if we are there any time at all. . . .

BEF General Hospital 18 Closes, U. S. Army Base Hospital 12 Departs

Letter # 137

Laura to Erle
Paris
Feb. 17
8 P.M.

Dearest Buddy: —

I'm on my way to Vannes which on the map is about four stations beyond Nantes. We suppose we are on our way home.

If I have an address there or stay any length of time I'll let you know. I do so hope you have your orders for sailing but I have no way of knowing.

We had our orders just two days before we started and I have written you two notes before this so hope you get some of them. I'm sending this by another 2nd Lieut.—my weakness. [***Erle related later that to mail this letter, Laura handed it from her troop train window to an American officer on another train stopped on an adjoining rail siding. Thus he learned that before embarkation she was to be stationed in a relatively nearby town.***]

I can't tell you how I hate to lose connection with you even for a few weeks but it can't be helped.. . . I go thru your town at 2 or 3 A.M.

Letter # 138

Laura to Erle
Vannes
Feb.19, 1919.

Dear Man: —

You have probably received beaucoup ***[many]*** letters from me telling you that I'm on my way. I wrote two or three times from Camiers and once on the train—the latter to be mailed at Nantes.

I wonder how your preparations for departure are coming along. But I believe that your preparations are complete and all you are waiting for now are orders. Have they come yet? We got ours Friday and we left Sunday.

Fifteen units have already gone and now the six that were with the B.E.F. are on their way. We all landed here unexpectedly but have been taken care of beautifully and the food—O Boy! You people who have been with the A.E.F. on A.E.F. rations are luckier than you'll ever know. I'm sure they all think here that we haven't had a square meal in twenty months.

Boarded up hospital wards

At first I had visions of running off to Nantes for a day to visit you but find it hardly practical. We have to report at all meals so if any orders come between times we'll be ready to start. I sort of have a hunch that we'll be sailing shortly. But you know my hunches are not the most authentic in the world.

We are out at Base Hospital 136 APO 935 which is situated in an old French barracks.

The town is very pretty—not so different from Nantes except much smaller and much fuller of Brittany costumes, etc. In fact today when I walked about the place I was quite homesick for you and our strolls about your city. . . .

Fifteen units have already gone and now the six that were with the B.E.F. are on their way. We all landed here unexpectedly but have been taken care of beautifully and the food—O Boy! You people who have been with the A.E.F. on A.E.F. rations are luckier than you'll ever know. I'm sure they all think here that we haven't had a square meal in twenty months.

Letter # 139

Laura to Erle
This letter was written on the inside of an envelope that had been pried open to provide space for a note. It was posted with a French stamp and "Received at Nantes February 22, 1919".
Vannes, Feb. 23, 1919 [Saturday]

Dear Man-o-Mine: —

I've just put in application for leave to come to Nantes for a day. I've asked for Tuesday [***26th***] so that if there is any chance of leaving soon I'll get in a small visit with you if it is possible. . . .

Letter # 140

Laura to Erle
Feb 24 1919 [Sunday]

Dear Buddy: —

Just received your letter of the 22nd. I didn't get my movement orders for Nantes yet but expect to get them so as to come Wed [***27th***].

Perhaps you'll get this in time. Anyway I'll hope to get you by phone when I arrive.

March-April 1919: Forsaken Angels

STOLEN MOMENTS

Letter # 141

Laura to Erle
Vannes, Mar. 3 1919

Dearest Buddy: —

I've just had your post card letter of the 27th and it was very welcome. You see we were all given Typhoid and Paratyphoid vaccines yesterday afternoon so we have all been miserable and cross and blue today. Your letter had a cheering effect.

Today one of our nurses had a note from a friend of hers in your unit saying that you all were going to Brest today. Now I don't know how late the information is but I'm hoping it is true and that I'll hear soon that you have sailed away to America. . . .

Our officers and men haven't yet put in an appearance and we've been here for two whole weeks.

My dear I've decided to never again complain because of my lack of conversational powers. Some of the conversationalists in here have nearly driven me to strong drink. But perhaps that isn't fair. One could hardly call a continuous chatter conversation.

So you didn't like the bottomless French beds. I love them. I fairly wallowed in mine last Tuesday night. Wed. night I didn't have time enough I was so busy listening to the clock strike. . . .

Letter # 142

Laura to Erle
Vannes, March 6 1919 [Wednesday]

Dearest Man:

Tonight is beautiful—moon, stars, balmy air and everything but the man. I've just come in from a movie at the red cross. We've been planning a trip out to St. Anne Auray. There is some sort of an annual religious pilgrimage and being that sort of a person of course I'll join the procession.

This afternoon we had a tea party and one of the girls put on her spink, spank, new, sparkly engagement ring by way of announcement. She is a beautiful girl—about the only beautiful person I've ever seen—and he is a very nice looking Englishman, a major in the M.G.C. [Machine Gun Corps] stationed for a long time in Camiers. After the party it was fun to sit and listen to the various ideas expressed about engagements, marriage, men, women, etc. . . .

Three of the units here have their movement orders to go to Brest but as yet we've nothing definite. We expect to follow them shortly altho we really know nothing about it. I imagine we'll know by Monday. . . .

As ever

Your Gertie -

a favorite quotation from Lt. Murray was

"Doity Goity Moiphy

She soitinly is a boid

She lives on thoity second

Right next to thoity thoid"

It is his conception of U.S. speech.

Wednesday I went down to Carnac and saw the dolmen (Druid stones, etc.) and a delightful beach for the children to play on some day.

Beaucoup ami or amou

or whatever the word is.

Private Baker's memoir:

March 8

On the way home!

We got packed up and I threw away all the extra junk. We actually left Camiers at 10:30 p.m. by way of 40 homme (men) and 8 chevaux (horses), or, "French Freight".

The Boston Unit, #5, boarded the same train at Boulogne, fIfteen miles away. We arrived in Paris at Gare du Nord (North Railway Station) at 1:00 a.m. and left again in twenty minutes. Arrived at Laval after a terrible night in the dinky boxcars in which 36 men were crowded, and we nearly suffocated. I spent my 27th birthday on this train. Two of our cars had warm axles and were condemned, hence the crowded cars. The French refused to give us more cars.

March 9

I got up at daylight, 6:00 a.m., as the train hit Laval. We beat it to a French canteen for coffee and we sure had to run in order not to miss the train. No one can tell how long a French train will remain standing still. I was crazy for exercise so I chased along the side of the car through the different stations at each stop.

We arrived at Redon at noon. Here the Boston and Cleveland Units went on to Vannes. Our orders were to proceed to Savenay, not far away. Upon our arrival at Savenay, we found our orders were wrong, but, we had to stay there overnight and were taken to the U.S. Hospital by real U.S. Army trucks.

March 10

Savenay, the largest U.S. Hospital center in France. There are about 10,000 beds here and the wounded are well taken care of.

At 10:00 a.m. we lined up and marched about fourteen miles for the train. Went back to Redon, the branch station, and on to Vannes where we should have gone in the first place.

March 11

Arrived in Vannes at 4:00 p.m., at the end of the third day in those small, crowded "petite" boxcars.

Hospital Unit #136 is the staff here. Our nurses were here also. We are in the French Cavalry Barracks located in large, well constructed stone buildings.

Letter # 143

Laura to Erle
March 13, 1919. [Wednesday]

My Buddy:

Just imagine getting a letter from you the very next day after you wrote it! This is twice such a thing has happened. How I hope your adj. [***adjutant: administrative officer or non-com***] was right and you'll get away toot sweet! We have heard nothing more except that there is no record either at St. Nazaire nor Brest of Hosp. 12—they even denied the existence of such an organization. [***Here the Base Hospital 12 nurses were again "forsaken," their records lost because they had been attached to a British hospital, forcing them to endure additional weeks of waiting for transportation back to the U. S. But the delay also provided Laura and Erle several unanticipated, much-treasured moments together in France.***]

Yesterday I walked over to Arradon and found it a delightful trip. If you come again to see me, as I hope you will if we stay long and if you [do], too, we'll spend it out there and along the road. Tuesday was such a happy, happy day—to me one of the very pleasantest we've had together. Since then I've done little but dream of it and of the future. And I'm so anxious to get started on our joining together. . . .

Today I've been thinking I'll have to tell Bertha and Mother of our plans. They'll be so hurt if I don't. And really it is such a little time away. Why it's no time at all till June.

I'm going to have a very busy spring and summer—learning to cook and sew and fixing up our home just so and taking care of you. Gee what big contract I'll have for the summer! One makes such contracts by the season doesn't one? or two? . . ..

Tuesday evening our men and officers came and such an excitement as there was when I arrived! No one had had time to miss me and no one knew I was late getting in except Emmy Matzen. The men have been told there will be no visiting with the nurses and they are certainly shy. They are afraid to say Hello for fear they'll be sent on to Germany.

It is a big disappointment to some of the girls.

We are having quite a sewing society in our room. The girls are making everything from baby dresses to tatting yokes and pink wool sweaters. I'm only an amateur and my handwork is rather a joke to everybody.

Tomorrow a bunch of us are going to Pont Chateau to a Calvary in sculpture. If it is worth while I'll tell you about it. I'm positively not going another place after that. I'm tired of sight-seeing. . . .

Letter # 144

Laura to Erle
Letter on reverse of three picture post cards of the Calvaire de Pontchateau, "Received at Nantes March 16-1919"
March 14 1919.

[*No salutation*]

Today a bunch of us went to Pont Chateau and walked out four kilos to see the Calvary. It was worth seeing.

There are the stations and the crucifixions all in life size figures. Then there are several groups, beardes [sic] and the Judgement. It is all in a lovely green park and is worth seeing. If you have a spare day and nothing to do I believe you'd enjoy seeing it all.

Anyway it is another day gone by. . . .

Letter #145

Laura to Erle
March 15 1919

My Own Buddy: —

. . . They've certainly shut down on all intercourse between the nurses and enlisted men. The C.O. has promised any man found being friendly with us to be sent to the 3rd Army. I'm surely glad we met no C.O. on our day together. That would have been a tragedy. . . .

Letter # 146

Laura to Erle
Vannes
March 19 1919

Dear Buddy: —

Yesterday I had your letter of March 16 telling about your offer from the W.N.U. man and about your unit not sailing and about our delightful day together. I did so want to sit right down and answer it but I was entirely too popular. The whole twenty decided to sit on or beside my bed and visit with me while they crocetted or knitted and embroidered and today has been so rainy that every one was forced to stay in and I've been nearly distracted. Finally they've mostly gone to movies and the final few are silent!

At this touching moment the gang has arrived and what do you think? There is a little balcony at the Red Cross movies reserved for nurses and officers. The nurses were sitting in the two front rows of it when the Colonel and a major came in with two mademoiselles and had a man come over and request four of the nurses to get up and give him their seats. When the request came in Miss Spencer said in a quiet ladylike way "The gentleman desires to sit. I think it wise that we all go and let them have our seats." And they all rose and filed out leaving the Colonel in possession. Just like that!

And now the babble!!! Ye Gods! And Hurrah for Miss Spencer! I can forgive her for a lot for standing up as she did. She has had a hard time of it from the beginning and how she hates it all! . . .

One of the units here (men and officers) are going to Brest tomorrow. So perhaps our time will come some day if we have patience. . . .

Letter # 147

Laura to Erle
Vannes
March 20, 1919.

My Dear

Today I went to see this [referring to picture on opposite side and other enclosures] beautiful chateau at Josselin. It is supposed to be one of the best examples of Gothic architecture in Brittany. But most of all I enjoyed the view from the ramparts, looking across the canal up toward the hills. . . .

Hospital #4 (officers and men) left today for Brest. Our men and the two other units waiting go tomorrow and the next day. I wonder if you got your orders.

A rumor came up today that we (the nurses) were to leave tomorrow at 3 P.M. but it was only a rumor. Our men's departure is encouraging however. It seems that they haven't forgotten about the unit completely—even tho we've been here four and a half weeks. . . .

This army business is maddening and yet this is the first time I've been kept waiting. You've had a much worse experience. I wonder at your patience.

Won't it be a G-r-rand and G-glorious Feeling to be one's own boss again!!!! . . . .

This would be an excellent camp if it did not rain so much. We have to line up in the mud and rain for fifteen minutes before each meal.

Private Baker's memoir:

March 22

I walked way out in the country and tried to talk French to the local Bretons but they "no compree"-did not understand much. This is Brittany and the Bretons do not understand good French. They are different from other French in some ways. They have their own dialect of French.

Got orders to leave for Brest after about two weeks here. We got up at 5:00 a.m. and did some fast packing and eating and then marched off at 7:00 a.m. Our unit nurses were all up and watching. We waited at the station until 9:00 a.m. and when the train came in it was a third class passenger train—swell for the troops

Although only about 110 miles, it was an awful ride to Brest. It took fifteen and a half hours, and say, you should have heard us sob and crab. I got about as sour as a peck of dill pickles. We only had a couple sandwiches during the period between 6:00 a.m. and 2:00 a.m. the next morning,

Upon our arrival at Brest we were given a big feed at the U. S. Embarkation Mess Hall. They can feed about 40,000 men in one hour here. After eating we marched about four and a half miles in the mud to the U.S. camp, Pontanagen. Gee, what a hike—all uphill. We were about dead from the long tedious train trip, loss of sleep, etc., before even starting up that hill so imagine how we felt when we got there. We reached camp at 3:30 a.m. and were put six in a tent—large square bottomed bell tents. There were spring beds for each man but no mattresses.

This would be an excellent camp if it did not rain so much. We have to line up in the mud and rain for fifteen minutes before each meal.

Our unit got orders Monday afternoon to be at the Embarkation Officers' Inspection Building at 7:00 p.m. for our honest-to-God last inspection in France. We passed, as usual. I was put on K.P. duty peeling spuds all night so I got little sleep. All men with names beginning with A, B and C were the goats. Caught an awful cold in the head.

The Nurses Again Forsaken

Letter # 148

Laura to Erle
Vannes March 23, 1919.

My Dear: —

. . . Yesterday I ran away to St. Nazaire. It is forbidden but I went anyway. It isn't much of a town but the blue, blue ocean looked well. I could almost see Lady Liberty on the opposite shore.

Our officers and men left yesterday for Brest. All the officers and men of the units that were B.E.F. have now gone and all the nurses except Unit 21 and ours. I don't know why we are left. I hope the shake up will help a bit but I'm almost resigned now. But my clothes are wearing out and I want to go home. (Loud splashing of tears.)

I wonder why they don't send you folks from Nantes. Doesn't anyone embark there?

As to another visit to Nantes after April 1st: I can't come either till after payday and that was on the 4th this month. I'm not sure either that I can get more than a twenty four hour pass but I'll try for a forty eight hour one if we are here that long. I can get a half dozen girls to come along with me if necessary but you'll have to promise not to kiss me when you meet me!! It is too embarrassing in public places for your little friend. (Did you think I left you abruptly on the corner that last evening in Vannes? I was afraid you would kiss me.)

If we both stay that long and if I can get the pass suppose you meet the noon train on April 5th and see what climbs off it. That would give us the afternoon and evening and the next day till 3 P.M. I'll get Frances to come with me and the shops will keep her busy. She was quite disappointed at not getting to go with me the last time. Cities have much more attraction for her than anything else. Our own C.O. has departed. He wouldn't give passes for any place for more than one day. I'm pretty sure the C.O. of the centre will let us go unless a fresh order comes. . . .

Since all the stray officers and men have left here there isn't half the danger of being taken up here for the infringement of rules. But we are freer in Nantes than Vannes and I'm absolutely bored stiff with this place and its surroundings. All French villages seem just alike. . . .

I'm not sure either that I can get more than a twenty four hour pass but I'll try for a forty eight hour one if we are here that long. I can get a half dozen girls to come along with me if necessary but you'll have to promise not to kiss me when you meet me!! It is too embarrassing in public places for your little friend. (Did you think I left you abruptly on the corner that last evening in Vannes? I was afraid you would kiss me.)

Letter # 149

Laura to Erle
Vannes
March 25 1919

Man-o-Mine: —

. . . We've heard nothing at all but there seems to be a general atmosphere of expectation of orders. Today the girls who went to Nantes returned with the news item that five hundred nurses were to sail April 1st. We've no way of ascertaining whether we are to be included in that or not but perhaps we'll be lucky.

However, I'm going right along planning to come to Nantes on April 4th—if I don't hear you've orders or if we don't have orders. So be on the reception committee please.

I want to see the Art Museum there if I come. The room here is full of raving about its beautiful pictures. . . .

Private Baker's memoir:

March 26

We started for the docks about five miles away at 11:00 a.m., arriving there at 12:30. We were fed a lot of buns, doughnuts and coffee on the pier by the U.S. Red Cross women.

About 4:30 p.m. we got in motion for our last time on foreign soil and we marched on board a U.S. tender and were taken out to the great ship, "Leviathan," which we actually boarded about 6:00 p.m. I was feeling pretty rotten all this time from my cold so had a good start for real seasickness. ,

Our unit was separated and we were assigned to our bunks immediately. I was the unlucky cuss to get in the worst group and was put in a compartment in the rear lower part, three decks below the main one, where the noise and roar of the propellers drives one crazy. I slept here three nights. I was sick and weak because I could eat nothing. To make things worse, the mess hall was terrifically hot and with a couple thousand going through it [at] all times, what little eats I did go after were spoiled.

It took about three days and nights to load this ship with its 15,000 human beings and its cargo. This [was] the most ever carried on one ship and the officers were worried. Feeding 15,000 people two meals a day is no small task and represents a marvel in efficiency.

Brest is a fairly large city and it looked fine from the harbor. There were a few U.S. destroyers in the harbour and a captive balloon.

Hurrah! Our ship actually started for dear old U.S.A. at 6:30 p.m. Everyone was on deck to see us go down the harbour and, boy, how we did leave old France behind. We passed the great capes and the lighthouses while land grew farther and farther away until finally it faded from view. It took just one hour and twenty-five minutes to be completely out of sight of France and the lighthouses. The old boat is now making 18 knots although it can do 26 or 28. The sunset in the west was gorgeous. The [sun] was right in our homeward path and, gee, what a feeling, 3,000 miles of water ahead. We were so happy we could hardly talk.

The six days that we were racing across the ocean were calm. I dreamed a lot about icebergs.

Letter # 150

Laura to Erle
Vannes, March 27

Dear Heart-o-mine: —

. . . So far we've had nothing in the way of orders. The expectation which seemed to be quite general has not yet died out but I'm dubious. I sometimes think that Uncle Sam has forgotten than [that] we are part of his family. The men of our unit sailed Tuesday and the officers today on the Leviathan. . . .

Letter #151

Laura to Erle
Vannes, March 28, 1919

Dear Buddy: -

. . . We are still hoping to have our orders to sail but nothing much to bank upon except a rather vague promise to get us away with the first bunch of nurses. But, my deah, while we wait I am so busy. I've made almost a whole lunch set the last three days. At least all the hemstitching is done and the corners basted on the cloth ready to [put] button holes on. Compree? [WWI American corruption of ***comprenez-vous***: do you understand] It looks coome sa [***comme ça***: thus] [***drawings of square table cloth and six square napkins***] Ain't they beautiful!! The little squares [in each corner] are Venetian lace.

And would you like to see what the dresser scarfs look like? The first one is lace bordered [***drawing of long rectangle with edging and two diamond-shaped insets***]. The left hand lace square is set on so poorly that I'll always have to keep a mirror or a pincushion or something over it to hide the stitches.

The second is better. It is hemstitched [***drawing of elongated rectangle with small squares inset in each corner and circles at each end***]. Aren't you thrilled over the symmetry of their lines?

It is funny here the way we all do fancy work. One day we all make dresser scarfs, another day it is little fancy aprons, another it is sweaters and finally luncheon sets. We are thinking of having an exhibition of our handiwork. Some of the girls do beautiful work and you know we've been here six weeks—almost—and absolutely nothing to do.

I've had two letters from home this week both written in March. We have a new Jersey cow and two baby chickens added to our family. I'm not going to take them lingerie. They have arrived too late to draw anything but some Clumy [*?*] lace to make up themselves.

Our unit was separated and we were assigned to our bunks immediately. I was the unlucky cuss to get in the worst group and was put in a compartment in the rear lower part, three decks below the main one, where the noise and roar of the propellers drives one crazy. I slept here three nights. I was sick and weak because I could eat nothing. To make things worse, the mess hall was terrifically hot and with a couple thousand going through it [at] all times, what little eats I did go after were spoiled.

We had fifteen days at Camp Mills but most of the time was spent in New York City. We were entertained with high class vaudeville and other shows, free, every day at the K. of C. Hut in the middle of Broadway and I was served with coffee, candy and cookies by Mrs. W. K. Vanderbilt and daughter, Muriel.

Letter # 152

Laura to Erle
Vannes
March 29

Dearest:

This evening came your letter of the 27th saying that your orders had come and you are busy on the home destination list. I'm so glad. I hope you really embark by April 5th.

We seem out of luck here and yet we are all ready and could easily go on 12 hrs. notice. So we may get orders at any time. . . .

Letter # 153

Laura to Erle
Vannes, March 31 1919

My Dear: —

. . . You weren't here so I took Emmy and walked out to Conleau, around the tiny islet and back to Vannes. We were gone just two hours and a half and didn't hurry a bit. The sky was so blue, the sun so bright and the clouds so fluffy.

Are you terrifically busy getting ready to depart? I wish I could be sure of it. I'm so discouraged yesterday and today that if it weren't for fear of brick bats I'd go out back of the barracks and howl. It seems that the powers that be have simply forgotten that we exist. We've heard that officially we sailed some time ago so are not on the sailing lists at all. And this hospital outfit here who sailed after the Armistice are signed up to go within two months. Probably they'll turn over this dump to us! That's the way I feel tonight anyway.

Our officers and men sailed on the Leviathan last Wednesday so should land about today or tomorrow.

The St. Louis unit who left New York the same day we did is waiting down at Carnac and has been there two weeks longer than we've been here. . . .

We'll probably be paid tomorrow and if I don't hear that you are departing I think I'll come down to Nantes about Friday. I know it isn't wise but I'm so sick of this place and I want to see you and talk awhile. I don't even know if I can get a pass but I think I can.

I'm going to bring Frances Hampton along this time. She was quite disappointed at not getting there before with me. I hope somebody else will want to come so I won't feel so much like a deserter. . . .

Private Baker's memoir:

April 2

Coming into New York harbour I nearly froze because I was out on the front deck early so as to see land come into sight. I had a camera to use on anything of interest.

We were all anxiously awaiting the first sight of the Statue of Liberty and it seemed to take a month for us to get close enough to see it. When we passed her we all gave her some lusty cheers. It was just 22-1/2 months today that I last saw her.

We landed at 9:45 a.m. after having viewed some impressive sights while coming into the harbor and up to our pier. All the great buildings were filled with people cheering and throwing colored streamers and ticker paper out of the windows. Factory and boat whistles screamed for an hour in their noisy welcome. It took about four hours to march all the troops off the great ship. When all the troops and the cargo were off, the ship rose about twenty feet. It stands three stories to the main deck.

April 3

I actually set foot on U.S. soil at noon and what a wonderful sensation. Home at last. It seems like a dream. We had to hang around the dock and were fed by the dear Red Cross women. I sent a cable home from the ship.

We took a ferry to Long Island for Camp Mills but our train didn't come for six or eight hours. We actually got to Camp Mills at 10:15 p.m. and everyone was about dead from exhaustion and lack of food.

We had fifteen days at Camp Mills but most of the time was spent in New York City. We were entertained with high class vaudeville and other shows, free, every day at the K. of C. Hut in the middle of Broadway and I was served with coffee, candy and cookies by Mrs. W. K. Vanderbilt and daughter, Muriel. Our meals were served free by the blessed women of Illinois at the "Governor Lowden Welcome Club." We had a free auto ride around New York, too.

Letter # 154

Laura to Erle
Vannes, April 5 1919

My Dear:

Orders have come!!!!!

At 5:30 this evening Miss Spencer made the announcement. We leave for Brest Tuesday the 8th. Did you get orders, too? I hope so.

Your telegram has not yet come but the letter with clippings was under my pillow.

I'll write again tomorrow when I'm a bit more calm but I want to get this off in the morning so you'll hear even if you leave Tuesday, too, as I hope you will.

Bobby was quote peeved at me because I didn't introduce you when we met them in the park!! [***Indicating, perhaps, that Erle had been able to make another trip to Vannes from Nantes.***]

Letter # 155

Laura to Erle
Sunday Evening
April 6, 1919.

Dear Buddy: —

I had your telegram this afternoon saying that your outfit leaves tomorrow or the next day and I'm so delighted. Maybe we'll go on the same boat!!!

I doubt if you will get this but I'll send it along anyway, hoping it will follow to Brest or even aboard.

Wasn't it lucky that I came to Nantes just when I did? How things are working out our way! If they'll just go on and our other plans go along so well.

This afternoon I went out to Quiberon by motor. I've motored pretty well over Brittany and Picardy too. The beach was so delightful. I've never been on a beach anytime when the air was so balmy. I know you'd have enjoyed it much even to the little shell fish that we played with. I saw a big jelly fish, too, and picked up little shells for the children [***her brother Silas's***].

Their Returning Glory Missed

Private Baker's memoir:

April 17

Got orders to leave camp for the West at 1:00 p.m., and we got started on time. We went directly through to Chicago without changing cars. We went through the Brooklyn tubes, the Pennsylvania tubes and over the D.L. & W. (Delaware Lackawana & Western) to Buffalo. Left Buffalo over the M.C. (Michigan Central) at 9:00 a.m. on the 18th and hit Chicago at 1:00 a.m., April 19th. We slept in the cars.

Headlines in Chicago newspapers:

STUDENTS HOME
FROM SERVICE
TODAY
Thousands On Hand to Welcome
Base Hospital units 12 and
13 From Overseas

BASE HOSPITAL UNITS
TREAD VICTORY WAY
IN TRIUMPHAL MARCH

UNIVERSITY HEROES, FRESH FROM
OVERSEAS, SWARM TO KHAKI AND BLUE CLUB AND RED CROSS CANTEENS, CHEERED BY WELCOMING CROWDS; STUDENTS JOIN IN

CITY GIVES WARM
WELCOME TO 300
MORE VETERANS

HUGS, KISSES, TEARS
HAIL HOSPITAL UNITS
WAR HEROES MARCH THROUGH
LINES OF WAVING FLAGS AND
CHEERING CROWDS

Subhead and paragraph within the story:

Seventy Nurses Left Behind

Seventy nurses who were attached to Unit 12 were left behind at Vannes, France. Some disappointment was expressed that these heroic women could not have returned with the company and share in to-day's welcome festivities…

Private Baker's memoir:

April 19

Got up at 7:00 a.m., and then our great homecoming started. Met brother Harold at I.C. (*Illinois Central*) depot as I got off the train. I was sure glad to see him. We marched to the Khaki and Blue Club for breakfast and dance from 8:30 to 11:00 a.m. We then marched through the city to the Hotel Lassen where we had lunch which was followed by speaking and singing. The noon crowds were all out to see us. Cheers and the waving of flags were the big features and we felt pretty proud for we had been overseas much longer than any other men around.

April 22

Same thing. System of demobilization is terribly rotten. All hoping we can get out today.

Last hour in the Army.

Lined up and marched to station at 6:10 p.m. to receive our final papers and money ($60.00). After standing for two and a half hours I got mine at ten minutes of nine in the evening. I was now a free man after close to two years of service, and I left for Chicago.

Excerpt from Memoir of Reserve Nurse Elizabeth Lyon:

[After 24 nurses had been detached from Base Hospital 12 to serve with the Army of Occupation in Germany]...The remaining nurses left Camiers on February 18, 1919 for Vannes where they spent seven weeks waiting for transportation home. April 7th they went to Camp Kerhuon by hospital train and April 14th aboard the *Prinz Frederick Wilhelm* at Brest, France, sailing for America the next day. They arrived in New York April 25th, and waited there to march in the Fifth [*Avenue*] Victory Loan Parade May 3rd, leaving for their respective homes the next day.

Their officers and enlisted personnel had earlier been feted in the unit's hometown of Chicago—an honor the nurses were denied—and in most cases the men already had been released from the Army.

Private Baker's memoir:

April 20, Easter

We had to lay around idle waiting to "sign here" and turn in stuff. A beautiful day went to waste. Oh, how we wanted to be home.

April 21

Signed papers and stuck close to our barracks. Nothing to do but wait and we have just spent five months doing that.

Letter from Reserve Army Nurse Mrs. Carrie Krost

HOTEL ENDICOTT

Columbus Avenue 81st to 82nd Sts

Manger Popular Hotels

Rates for Hotel Endicott

Rooms $1.00 With Bath $1.50 & $2.00

New York, April 29, 1919

Dear Mother:

We are still here. Gerald [*Capt. Gerald Krost, MD*] is out of the Army and is leaving for Chicago tonite.

I think I will be able to leave here Saturday or Sunday but Saturday I think.

Had my physical exam yesterday and was found O.K.

We are pretty busy seeing New York but the side walks do hurt our feet I must say. . . .

You can send my letters to 124 N. Main St. Crown Point, Indiana for that is the place we are getting cleaned up and I am sure dirty.

My suit is awful but as it is all I have to wear I can't very well get it cleaned. And as I am not out of the Army until about the sixth of June I have some little time to wear it. I have thirty-six days leave coming to me. . . .

[From unidentified Chicago newspaper, April 1919]

NURSES ANGELS

OF THE WOUNDED,

COL. BESLEY SAYS

Col. Fredric A. Besley of Chicago, commander of base hospital unit No. 12, under whose direction thousands of American wounded have been treated, is home from the war.

Col. Besley, fortified with an honorable discharge, arrived in Chicago last night.

"Who won the war?" he was asked last night.

"Well, we all won the war," he answered with a laugh." "But in this regard I want to give the fullest measure of praise to those brave women of the United States and Great Britain who acted as nurses in France.

"Barred from a zone within seven miles of the front, they worked their way up to the limit allowed, and there, night and day, often for long periods without sleep, they encountered and overcame every trial that presented itself. They were the angels of the wounded American soldiers.

"Their bravery was evidenced on May 19, 1918, when the Hun bombed our hospitals. There were over 500 new casualties as a result of the raid but not a woman flinched from her work. They went around quietly amid the crashing of bombs, caring for the old cases and the fresh ones alike. It was wonderful."

Col. Besley declared the influenza epidemic was severe in the overseas camps and the number afflicted large but the fine physical condition of the men seemed to keep the death rate below that in the United States. He was accompanied home by Lt. Col. Milton Mandel.

With these words, Dr. Besley here redeemed himself publicly from the privately expressed ire of the nurses against him.

Extract from The Medical History of the United States Army in the World War:

During its active service with the British Expeditionary Force, Base Hospital No. 12 cared for 27,438 British and 2,229 American medical cases; for 30,010 British and 966 American surgical cases.

The 60,643 total cases did not include the hundreds, perhaps thousands, treated at the Casualty Clearing Stations by numerous surgical teams sent to the Front by Base Hospital 12, especially in the last three months of hostilities. In his official report in January 1919, Major Payson L. Nussbaum, who commanded Unit 12 after October 9, 1918, stated that 29,795 patients were evacuated to England, 19,784 sent to Convalescent Camp and 7,322 returned to duty. This would leave 3,742 unaccounted for, presumably dead. However, he also gave a "percentage of death rate to total admissions" which calculated to 4,621 deaths. Note that there did not seem to be consistency in statistical reporting from one unit to another.

Letter to Miss Harriet Davis

This, the final letter in the long series, came in a regular stationary envelope with an Army censor stamp, using American Red Cross post cards for the message and marked "Received at St. Nazaire April 9, 1919".
412 E. 15th St., Indianapolis
North Vernon, Ind.

May 18, 1919

Dear Harriett :

Thanks for your most cordial invitation to come up and get acquainted. I'll surely accept when I come to your city to do some necessary shopping and to visit the Allens.

Of course I know Audrey but I haven't seen her for a very long time. And you were just a little girl when I saw you last. So I'll just have to get acquainted again with you all.

I had a delightful little visit with Mrs. Rees and Erle this week. But of course he has told you all about it.

I'll warn you when I expect to come to Indianapolis and I'll be much disappointed if you don't ask questions. I won't promise to always know all the answers but at times I have a sparkling imagination so will do my best.

Sincerely,

Laura H.

Laura in Louisville, after France.

Where we fought
the war Base 12
65

VII EPILOGUE

LAURA G. HUCKLEBERRY AND JOHN ERLE DAVIS

By prearrangement, Laura and Erle, both still in uniform, reunited in Louisville in May 1919. Each was discharged from the U. S. Army within weeks.

Former U. S. Army Reserve Nurse Laura G. Huckleberry married former U. S. Army Sergeant First Class John Erle Davis in the rectory of Sacred Heart Roman Catholic Church, Washington, DC, on September 19, 1919. Her brother Will Huckleberry, a War Department clerk, and his wife witnessed the ceremony. It had been necessary for Erle, who was covering the Senate for the *Washington Star,* to obtain a dispensation from James Cardinal Gibbons, archbishop of Baltimore and a distant cousin, in order to marry Laura, a Methodist.

They were both proud of their service in France and became active in the American Legion, an organization of World War veterans.

Laura never practiced the nursing profession again except to care informally for ill family members. She was active in the League of Women Voters, a legacy of her suffragette experiences. Erle was an editor and columnist for two different Louisville newspapers, and joined a fellow journalist in 1923 to found a successful Louisville advertising agency.

In 1947, a "Red Cross girl" from his base hospital unit, who sought permission to use his name in a novel she had inspired, contacted Erle. In the first edition of *Came a Cavalier*, by well-known romance history novelist Frances Parkinson Keyes, a minor character is Sgt. Davis.

Laura and Erle never returned to Chicago except on business trips, nor did they ever visit England and France again. Erle sold his advertising agency to his employees in 1953 and they retired to Florida.

They produced three children, John Lehigh, born 1920; Ann McCarroll, born 1922, and Michael Warren Rees, born 1931. John died a month before his 14th birthday of scarlet fever and strep throat. In 2011, three of Laura and Erle's seven grandchildren were active in the medical field. Jane Skubik Dupperault was an administrator at Stanford University Medical School, Harriet Skubik Hanley was a health issues expert for the California State Government, and John Mattison Davis, MD, was a board-certified psychiatrist in Louisville. Dr. Davis served nine years active duty, 1999-2008, in the US Navy Medical Corps, including two deployments to Iraq and two years in Okinawa with the Marines.

Laura was hospitalized in Sarasota, Florida, in 1957 with dementia from acute arterial sclerosis and died October 31, 1962, at the Veterans Administration Hospital, Perry Point, Maryland, two months short of her 75th birthday. Erle died January 8, 1985 in Pontiac, Michigan, in his 97th year. His last by-lined newspaper story was published in the *Detroit Free Press* when he was 95.

GEORGE R. BAKER

Private George R. Baker, born March 8, 1892, of Rock Island, Illinois, was a student at Chicago's YMCA College when he joined the Northwestern Unit in May 1917. He was promoted to Private First Class before release from active duty in 1919. He returned to his hometown where he married Thekla Heimbeck in 1921 and became a prosperous life insurance agent and was active in the American Legion, for which he served as the local post commander. He also was an active participant in Base Hospital 12's annual reunions in Chicago.

Following World War I, George completed his enlisted military reserve time by being commissioned in the early 1920s as a second lieutenant, Cavalry, U. S. Army. After the United States was drawn into the Second World War, early in 1942 he was recalled to active duty in the U. S. Army Quartermaster Corps and served in several posts around the United States before being honorably discharged as a first lieutenant at the age of 52.

George and Thekla produced one child, George Jr., born 1931. George Baker Sr. retired in 1960 after 35 years with Mutual Benefit Life Insurance Company. He died of heart disease in October 1962, the same month as Laura Huckleberry Davis's death.

OTHER PERSONNEL

Daisy D. Urch, ITS '13, the first chief nurse of BH 12 who enlisted the original crew of nurses, largely from Cook County Hospital and the affiliated Illinois Training School for Nurses. Born in rural Oakland County, Michigan, June 4, 1876, she was 41 when she recruited her nurse crew for the Army. After teacher training, she had been a kindergarten teacher in Clarkston and Munising, Michigan, for nine years before entering ITS in 1910. Despite her evident personality conflict with Dr. Besley—who had recruited her to enlist the nurses—and with the Regular Army officer commanding BH 12, after she was transferred to a different U. S. Base Hospital in 1918, she received favorable ratings.

She was released from active duty September 30, 1919, somewhat later than the BH 12 nurses, and moved to New York City where she taught nursing at the City Hospital for a year and a half. Subsequently she earned a B.S. degree in nursing school administration from the Columbia University Teachers College in 1923, and then moved to California where she taught nursing at hospitals in San Francisco, Oakland, Alameda County and Los Angeles.

By 1934, she had moved to Minnesota where she was education director of the State Board of Nursing Education. Subsequently she was involved in nursing education in the Rochester, Minnesota, area until she retired in 1945. During the 1920s and 1930s, she published numerous articles on nursing education in *The American Journal of Nursing*. Miss Urch died June 10, 1952, at the age of 76.

Ruth Spencer, ITS '11, of Ludington, Michigan, succeeded Daisy Urch as chief nurse in December 1917. In the 1941 reunion directory of Base Hospital 12, she was listed as having died.

Isabelle E. Carruthers, hometown, Chicago, listed as assistant chief nurse at war's end. No further information.

THE TWO NURSES KILLED MAY 20, 1917, IN SHIPBOARD ACCIDENT:

Mrs. Edith Ayers, ITS '13, a 36-year-old widow from Attica, Ohio. Mrs. Ayers' early personal history is related in the biographical section at the beginning of this narrative. Her premature death had a high impact on the small community where she originated. When her body arrived Thursday, May 24 at the rail junction a few miles away in the next county, a delegation of 32 automobiles filled with Attica and Seneca Country residents drove over to escort the remains home for the funeral Saturday at the Methodist church. The delegation included town officials, representatives from the local brotherhood orders, friends, relatives and members of Company I, 8th Ohio Infantry awaiting Federal call-up. Eighty-nine years later, there still is a memorial display about Mrs. Ayers in the town museum.

Helen Wood of the Evanston Hospital, a 28-year-old native of Scotland who had come to the U.S. only a couple of years before. Two of her brothers were serving in British forces in France, and it was reported, but not verified, that both were killed in action. Her military funeral in the northern Chicago suburb was held a week after her death. An escort composed of members of the Chicago chapter of the Red Cross and the Illinois medical reserve corps and fifty sailors from the Great Lakes Naval Training Center met her body at the Chicago railroad station. At the services at the First Presbyterian Church of Evanston, the procession included doctors and nurses from the Evanston Hospital, all in their white uniforms, and then the others, accompanied by the GLNTC band.

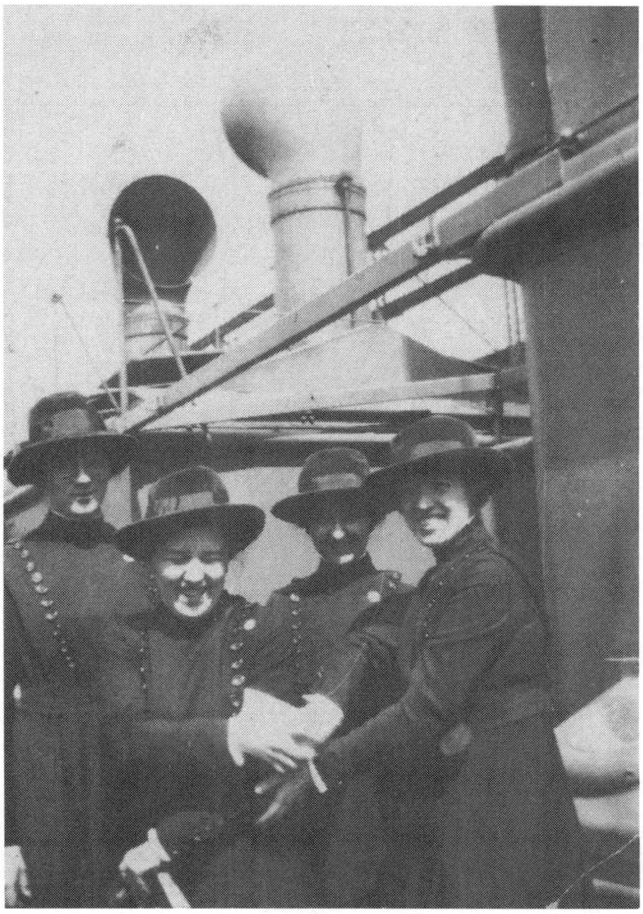

Unnamed nurses on a ship.

TWO NURSES WHO DIED OF DISEASE WHILE IN FRANCE:

Florence A. Hinton of Decatur and Urbana, Illinois, a 1914 graduate of ITS, head nurse of the "head ward" and one of the original BH 12 nurses, died of meningitis, January 20, 1918, after working her ward until 4 p.m. the previous day. She was interred in the British military cemetery at Etaples. An historical article about her in the Decatur *Herald & Review* noted that Miss Hinton was born in Leeds, Yorkshire, England, November 13, 1879, and came to the United States at the age of 10 years. At the time of her death, her family had lived in Decatur 23 years, having lived previously in New York and Pennsylvania. Hinton attended high school in Oneonta, N.Y. After graduating from ITS, she became supervisor of surgical wards in the Akron, Ohio, city hospital for two years and did private nursing in Champaign, Illinois, from September 1916 until May 1917, when she left for France. She was a registered nurse in the state of Illinois. She was a member of St. John's Episcopal Church in Decatur, where she was active in the altar society and the choir.

Lucile Pepoon, hometown Chicago, died of disease in France November 29, 1918. She had contracted "trench fever" in the spring of 1918, and was hospitalized from then to her death, which ultimately may have resulted from flu. Trench fever was a rare disease that appeared during the war and was attributed to infection spread by lice. It was characterized by sudden fever, headaches, malaise, and lasting pain, especially in the shins. While debilitating, it was not known to be fatal. Miss Pepoon was the daughter of a Chicago physician and was doing school-nursing when she left for France. She also was buried in the English Military Cemetery at Etaples—near the soldiers both nurses had so faithfully nursed.

FOUR NURSES WHO WERE INVALIDED HOME:

Gertrude Lambert, hometown Evanston, IL, invalided home

Belletta Paulson, ITS '14, hometown Granton, WI, invalided home

Margaret Powers, ITS '07, hometown Marinette, WI, invalided home

Frances Sherman, invalided home; no other information

LAURA HUCKLEBERRY'S OTHER PRE-WAR APARTMENT ROOMMATES FROM 453 BELMONT, CHICAGO:

May (Conny) Connard Alexander, ITS '13, born March 24, 1889, in Elwin, Illinois, listed her hometown as Decatur, Illinois. She was one of the youngest of the nurses and also the most adventurous.

Originally one of the group of ITS classmates sharing a northside apartment in Chicago, she left early as a volunteer with a Red Cross group in Germany in 1916, and did not join BH 12 in France until October 1917 in the first wave of reinforcements. In October 1918, she contracted tropical dysentery, and was confined for nine weeks in British hospitals in France. When she returned to BH 12 in December, it was starting the process of closing.

Back in Chicago, she returned briefly to Cook County Hospital as head nurse of the women's surgical obstetrical ward, and then took a job as a public health nurse in Montrose, Colorado, a remote mining and ranching area. She then exercised her veteran's right to obtain a homestead property of 640 acres in the mountainous sheepherding country east of Montrose, learning how to ride a horse and hiring help to establish the property so that it could become legally hers.

After a couple of years of this mixture of public health nurse and outdoor life, she returned to Chicago seeking a nursing position and ended up in Oklahoma as a rural nurse for the Tuberculosis Association, where she learned to drive a Model T Ford to make her rounds.

Finally she returned to Chicago, married a Mr. Alexander, and settled down to raise a family. According to the 1941 BH 12 reunion directory, she lived in Peoria, Illinois, at that time. In 1977, she prepared a memoir for her grandchildren on her experiences, which was published in several ITS alumni newsletters.

Bess Gambee, ITS '13, born January 20, 1885, in Attica, Ohio, same as **Edith Ayers**, who was her cousin. After World War I, she was a public health nurse in the Chicago area, in 1967 living in Oak Park, Illinois She died November 4, 1984, at the age of 99 and is buried in the Attica Venice Cemetery, Seneca County, Ohio.

Frances Hampton, ITS '13, born about 1881 in Cattlettsburg, Ky., her hometown in 1917 was Frankfort, Ky. Following the war she married a Canadian officer, **Lt. Duncan McCosh**, afterwards lived in Toronto, ONT, and had no children. In 1930-

31, when the Davis family lived in Buffalo, New York, they frequently visited with the McCoshes in Toronto. One lure was that Canada did not have Prohibition against alcohol, as the U.S. had had since 1919. Laura corresponded regularly with Frances. Frances died in 1955 and is buried alongside her husband, who died in 1973, in his hometown cemetery at Paris, Ontario.

Edith (Mac) McMillan (also spelled McMillin in some records), ITS '13, whose hometown was Chicago, married a former patient, U.S. Army Capt. D. L. DeCant in August 1919. He had been wounded in the battle of Soissons. In 1941, she was listed in the Base Hospital 12 reunion directory as Mrs. Clarence Peterson.

Emma Matzen, ITS '13, born September 12, 1879 in Columbus, NE, had a daughter after the war who also became a nurse. In later years, she lived in Ft. Wayne, IN near her daughter and corresponded as late as 1977 with Erle after Laura's 1962 death. She died December 1979 in Ft. Wayne.

Ella Pawlisch, ITS '13, hometown North Freedom, WI, married a Mr. Gibson postwar and in 1967 was the assistant postmaster in North Freedom.

OTHER NURSES:

Elizabeth T. Cleveland, Laura's roommate in France, like May Connard and Florence Hinton, she listed her home as Decatur, Illinois. She was born in Yorkville, Illinois, January 20 1880, and lived at 18th and California, Chicago, when she enlisted in BH 12. She was not a graduate of Illinois Training School for Nurses but had been active in Red Cross nursing for a number of years. She was one of the nurses who volunteered for duty in the Army of Occupation in Germany in February 1919. She returned to Chicago in September of that year and married Walter Morf in 1923. She had been secretary of the Chicago Red Cross Nursing Service when she died childless April 15, 1931 in Chicago.

Carrie G. Krost, hometown West Salem, WI, married pre-war to Dr. Gerard Krost, rare husband-and-wife team in same unit overseas. Died in 1963.

Gabrielle Horowitz, born October 28, 1882, and died in Chicago, December 5, 1981. No hometown reported, but according to Laura's letters, she was an immigrant from Eastern Europe, and was still living in Chicago at time of the unit's 1967 50th anniversary. Along with Bess Gambee, she attended several BH 12 reunions held in the Chicago area.

Boats on the English Channel.

Regular Army Col. Christopher C. Collins, who commanded the unit from May 8, 1917, to September 1, 1918, rose to be chief of Army Medicine before retiring, one of two former BH 12 officer so recognized. He was listed as having died prior to Unit 12's November 1941 reunion.

Capt. A. E. Magee, Regular Army adjutant of BH 12 when it embarked for England and France in 1917, also rose to become head of Army Medicine.

Col. Frederic A. Besley served as professor of surgery at Northwestern Medical School for 25 years. He died on August 16, 1944 at the age of 76, survived by his wife. He had served as the President, Secretary and Treasurer of the American College of Surgeons, of which he was a founding member and a Fellow. He also had a clinic in Waukegan, Illinois.

Capt. Gerard N. Krost returned to the Northwestern Medical School faculty, where he served until retiring in 1954 as emeritus professor. He also was affiliated with Wesley Hospital (later Northwestern Memorial) as a department head. After beginning in surgery he specialized in pediatrics. Dr. Krost was reported to be suffering from a broken vertebra at the time of the November 1958 BH 12 reunion. His wife Carrie died in 1963.

Lt. Col. Milton Mandel died August 28, 1950, at the age of 69. He had been affiliated with Mercy Hospital and was a faculty member at the Loyola University School of Medicine. Dr. Mandel was promoted from Base Hospital 12 to be "consultant of medicine" for the Second Army Corps in 1918.

Major Payson L. Nussbaum, whose specialty was gynecology, returned to the Northwestern Medical School where he served on the faculty until 1932. He died July 29, 1936, at age 57.

Major Walter H. Nadler, who held both undergraduate and medical degrees from Northwestern, was a member of the NU Medical School faculty from 1916 until his death from lung cancer on September 13, 1952. After the war, he also was affiliated with the Wesley, Passavant and Cook County hospitals and the Hines Veterans Hospital.

Dr. M. Payson Neal was Base Hospital 12's pathologist as a first lieutenant until his transfer early in 1918. From September 17 to October 30, 1918, he was hospitalized with influenza and pneumonia, during which period he was promoted to captain. In March 1919, he was advanced to major. He retired from the University of Missouri School of Medicine as professor of pathology.

Lt. Col. Kellogg Speed returned briefly to the Northwestern Medical School faculty and later became an internationally known orthopedic surgeon who contributed numerous articles to the medical literature. In March 1918, then-Major Speed married a nurse, **Margaret Rudd**, evidently from another hospital, since she does not appear on any of the BH 12 nurse rosters. When they returned to the U.S. in January 1919, they brought with them their baby daughter. Dr. Speed died in Chicago July 2, 1954, of heart disease, at which time the Journal of the American Medical association devoted a full page to his accomplishments.

ENLISTED PERSONNEL

Arthur R. Hall, Laura's favorite orderly, was promoted to sergeant before the unit returned to America. After the war, he returned to Northwestern Law School to obtain his law degree and became a prominent Chicago attorney specializing in corporate law. A resident of Evanston, he died at the age of 54 on February 10, 1948, survived by his wife. A former colleague wrote to the BH 12 Reunion group "so sorry to hear of the passing of Art Hall, wonder how many on the old 12th knew of the many, many members who had been aided in the past by Art's kindnesses and wisdom—I personally know of several."

Of the other Northwestern students and graduates who enlisted in BH 12 in 1917, four—**Fred Bollman, Wesley Busby, Herbert Roberts** and **Frank Silver**—gained commissions before war's end. At least eleven others besides Hall were made non-commissioned officers. And the enlisted ranks of BH 12 in general produced an amazing 16 future medical doctors and two dentists.

Outstanding among these was **Michael L. Mason**, born April 23, 1895, in Rossville, IL, who obtained his B. S. degree from Northwestern in 1916, and his M. A. in 1917. He was a sergeant first class in charge of orderlies by war's end, earned his M.D. degree from Northwestern in 1924 and a Ph.D. in surgery in 1931, to become a prominent Chicago surgeon as well as professor at Northwestern Medical School. Then in 1942, Dr. Mason, as a major, organized another Northwestern hospital unit as a World War II U.S. Army hospital unit of nearly the same designation, General Hospital 12, which served in North Africa and Italy with distinction. Dr. Mason was discharged in 1945 as a colonel and died March 30, 1963, survived by his wife of 42 years.

Norman Klein, the enlisted private in Base Hospital 12 who regularly supplied by-lined stories to the *Chicago Daily News* during the unit's overseas adventures, had been a student at Kansas University and University of Wisconsin. He was born in Kansas City, Missouri in 1897 and died in New York October 27, 1948. Although he had served as editor of *Base Talk*, BH 12's newspaper, Klein was not among the privates promoted by the end of 1917, not particularly remarkable since the "un-promoted" group also included some prominent Chicago family names of the time. After the war he settled in the New York City area where he was active in journalism, advertising and public relations. He wrote an article for the September 1940 issue of *Esquire* magazine about the 1917 incident on the *SS Mongolia* in which the two nurses were killed and one wounded.

OTHERS

It is not known whether **Lt. Robert Murray**, late of the Canadian Expeditionary Force, holder of medals for bravery and wounded in action at least twice, returned to Scotland or Canada after the war. A Lt. Col. Robert Murray was reported to have died in Scotland in August 1936, but it is unlikely that it was the same person.

Lt. Duncan Darragh McCosh, wounded and gassed veteran of the Canadian Army, died in 1973 and is buried alongside his wife, the former U. S. Army Reserve Nurse **Frances Hampton** of Kentucky who died in 1955. They are interred in the Paris, Ontario, cemetery.

Mary Blampied of Jersey, Channel Isles, Laura's V.A.D. in 1917, moved to Louisville, Kentucky, in the mid-1920s to be the private secretary to a wealthy woman, Mrs. Ballard of Ballard Mills, and thus became a close friend of the Davis family who lived in Louisville. Around 1935, she married a British Colonial Office civil servant and moved to Dar-es-Salaam, Tanganyika, British East Africa. The Davises were fascinated by the picture magazines and African artifacts she sent as gifts. After Laura's death, she continued to correspond with Erle and Michael. She died in the early 1960s on Jersey.

Chaplain Jesse S. Dancey returned to America after serving 14 months in France, lecturing about his experiences and generally supporting the nation's efforts. He urged America, prophetically, to bond with Great Britain post-war, rather than depending on its other Great War Allies.

VIII ACKNOWLEDGEMENTS

Army Medical Corps: Dr. John T. Greenwood

American Red Cross Archives: Jean Waldman Shulman

YMCA Archives

Salvation Army Archives

Chicago Historical Society

Chicago Public Library

New York Public Library

Canada Library and Archives

University of Wisconsin Archives

U.S. National Archives

Kansas University Archives

Northwestern University Archives: Patrick Quinn, Janet Olson, Kevin Leonard

Seneca East County Library, Ohio: J. Victoria Eckenrod, Director

Berkley, Michigan, Public Library: Celia Morse, former director

Daughter Elizabeth R. Davis and Niece Harriet (Hattie) Skubik Hanley

George Baker Jr., Rebel Kreklow, Joseph W. Oglesby Jr., Professor Emeritus Henry Hager, Professor Emeritus Charles K. Hyde and Professor Marian Moser Jones

Canadians Robert Lynch, Paul Fleet, Ted Douglas, Michael Beale

Hospital tents.

IX EDITORIAL NOTES, BIBLIOGRAPHY AND SOURCES

EDITORIAL NOTES

Throughout Laura's letters and the letters or memoirs of other nurses, doctors or a soldier of Base Hospital 12, there are some spelling and punctuation conventions which seem incorrect today but apparently were acceptable when they were recorded. For instance, the punctuation mark in these documents commonly was placed outside the quotation mark, a British convention but not now an American one. More noticeable are shortened spellings, for example, "thot" for "thought" and "c" as shorthand for the Latin "*cum*," meaning "with." The frequent use of "staid" for "stayed" may be another version of this simplification. There have been numerous largely unsuccessful literary movements over the decades in the U. S. for adoption of simplified spellings, and the aforementioned examples may have been typical of the First World War period in the American Midwest.

And there are instances of "there's" being used with plural objectives, grammatically wrong but certainly common in conversation even today. Laura also made occasional, but common errors in use of apostrophes, spelling of commonly misspelled words such as "judgment" and she confused "effect" with "affect." Considering that Laura's letters were written deliberately in "flow of consciousness" conversational style and often in dim light under trying wartime conditions with no dictionary handy—much less a computer's "Spelling" and "Grammar" check programs—her writing is remarkably literate for a small-town high school graduate with one year of college. Indeed, her well-educated descendants could learn from her precision.

BIBLIOGRAPHY AND SOURCES

Baker, George R, Jr., *Heroes and Angels*, 1999, Baltimore.

Barry, John M., *The Great Influenza*, 2004, Viking, New York.

Bishop, Alan and Mark Bostridge, ed, *Letters from a Lost Generation: First World War Letters of Vera Brittain and Four Friends*, London, 1998; Northeastern University Press, Boston, 1999.

Dock, Lavonia L. and others, *History of American Red Cross Nursing*, New York, Macmillan, 1922.

Dubois, Marguerite-Marie, *Larousse's French-English/English-French Dictionary*,1955, Pocket Books, New York.

Hart, Captain B. H. Liddell, *The Real War 1914-1918,* 1930, Little, Brown and Company, Boston, Toronto and London.

____________, *A History of the Illinois Training School for Nurses*, 1929, Chicago.

Holman, Louis A., *Hornby's Etchings of the Great War,* Boston, 1921, Goodspeed.

Keegan, John, *The First World War*, Vintage, New York, 2000.

Lusk, Brigid, "Progressive Era Women in Medicine," *Chicago History,* Chicago Historical Society, v. xxvii, No. 2, Winter 2000.

Macdonald, Lyn, *The Roses of No Man's Land*, Joseph, 1980, London.

Macpherson, Sir W. G., edited, *Medical Services of the Great War: General History,* His Majesty's Stationery Office (HMSO), London, 1921-24, v. 3, Medical services during the operations on the western front, 1916, 1917 and 1918.

McKinley, Albert E., Charles A. Coulomb and Armand J. Gerson, ed., *A School History of the Great War,* American Book Company, New York, 1919.

Schneider, Dorothy and Carl J., *Into The Breach: American Women Overseas in World War I*, Penguin, London, 1991.

The Editors of Life, *The First World War*, New York, 1965, Time, Inc.

War Department, *Official History of the War, Army Medical Corps*, Government Printing Office, 1923, Washington, DC.

X APPENDICES

1. DIARY OF RESERVE NURSE LAURA HUCKLEBERRY MAY 17-JUNE 13, 1917

May 17, 1917

12:15 am

On our way at last. A big crowd to see us off—Star Spangled Banner etc. Felt rather like a corpse, but am still able to tell the tale.

Frances, Mac and Bess are all in my car, also Miss Cleveland. Edith is ahead with Daisy and Ella & Emma are in the rear car. Our train is a special c only officers (MDs) and a few wives and the nurses.

PM 5-17-17

Up in time to stand on the bridge as we ferried across the Detroit River. Our meals are served table-des-hote and we pay 75 cents and are refunded it in NY. The meals are only fair.

At 1 o'clock we passed thru St. Thomas Ontario. A big crowd was down to the station to greet us. They brot us tiny Canadian & English flags, bananas, oranges, ice cream cones.

So many asked us to look out for their sons and brothers. They have a population of 19000 and have sent 2000 men to the front.

May 19

Forgot to write yesterday. We were breakfasted at the terminal on Hoboken Docks yesterday by the Red Cross and it surely was a failure as a breakfast. Then we went to the Red Cross quarters in the Waldorf. There were fitted out c hats soft, velour and other navy blue serge dresses, red cross capes, brassards & caps.

I like the brassards caps & capes very much. The latter are navy blue & lined c bright red flannel & a red cross on the front—nifty.

We had lunch at the Waldorf & finally escaped & took a bus ride up town. Saw Central Park, Riverside Drive, etc. Was able to recognize a few points from pictures, etc.

We went to dinner on the St. Louis. That is another boat that is taking Unit 10 from Philadelphia.

Then we were assigned our rooms on the Mongolia. I am with Miss Hostman and Miss Hoffman—neither is very interesting, but I can stand them a while at any rate.

Frances is in the next state room and Bess in the next.

Emma and Mac are the only two of our bunch together.

This morning we were not allowed to go off the boat at all except over to breakfast on the St. Louis.

The meal was fair but the service was rotten as most of the waiters were suffering from channell fever.

This boat is an old one and has been used as a freight but is fixed up now as a transport.

They're afraid to risk the good boats. This is the boat that sunk a submarine not long ago.

I understand that we are to be convoyed across but of course do not know.

We don't even know where we'll land but our destination is supposed to be France.

Later

We left the docks at 3 o'clock were towed down the channel & stopped just inside the harbor.

The gunners were practicing loading their guns & overheard them say that it was $600 to the man that got a submarine. Also that the Germans were offering 1500 to the man that got us.

The boys who had signed in as privates were all shooed off the upper decks & we all so hated to see them go. They are such nice boys & so willing to do anything.

While we were at dinner the boat started again and at 8:30 passed the Ambrose Channel Light out into open sea.

All the port holes are closed and no lights are allowed anyplace, even in the saloon nor corridors later.

We walked in the wind & listened to the boys playing their ukuleles & singing. They are much gayer down below than we are above.

The two meals we have had on board have been very good.

5-20-17

Awakened to find the air so nice & fresh and cool. Went up on deck before breakfast. Was the first lady to breakfast.

Pawly and I walked out on foreward deck & talked to the officer who was left behind yesterday & came after us in a tug. He told us several interesting things. Among them is that the gun on the stern is the Teddy Roosevelt and it was the first American gun fired in this war. The guns on the bow are called Martha and George Washington.

It is very quiet today—only a few white caps. We'll be sure to have rough weather before we reach England—so everyone says anyway.

6:30 pm

At 2 pm we had target practice. We watched them throw off the target & then the boat turned around. We watched them load and fire & then Emma said "Somebody's shot!"

I turned and saw 2 girls on the deck & blood all around. One was Helen Wood from Evanston. The other was Edith Ayres. Emma had a flesh wound in the hip & one in the arm.

Edith was shot in the temple and Miss Wood in the heart. Something was wrong with the gun. We might all have been killed but evidently our time has not yet come. I can't bear to think of Edith's people.

They have turned the boat & we are headed back for New York.

5-22-17

Yesterday morning we anchored and a tug met us & took Emma, Dr. Besley and Pawly & Miss Powers ashore. Emma smiled and waved to us. Her wounds will not be serious unless something unexpected developes.

Later we were towed up channel & anchored in the middle of the Hudson & no one is allowed shore leave—I suppose to guard against desertions. But I don't believe there is a nurse or doctor or rookie who would turn back if they could. However, there was some trouble in securing a crew and I suppose the company doesn't want to risk getting another.

The bodies were put in caskets, draped in flags & put on deck for a short service by the chaplin & then lowered over the side to a government tug & taken ashore. Bess and Miss Poole of Evanston were to go along but at the last no one was allowed to accompany them.

Polly got back about 5 o'clock. They were brot in the Admiral's launch.

They left Emma in the Naval hospital. The X Ray showed nothing but of course the wound is necrotic. She has been so brave about it all. Everyone speaks of her grit.

And we are all so glad that the wounds are no more serious.

Today is gray & chilly but we've been out on deck most of the time anyway. We've walked & talked & watched a bunch play shuffleboard. Then we all wrote letters to Emma. Poor girl, I know she hates to be left behind.

5-23-17

Last night we had a party—singing, reading and dancing & flirting. Some were very gay. I drew a mushy one but then it couldn't be helped.

We started at 5 this morning but had to anchor just outside the harbor till the fog cleared.

They have heard that a German raider is about & also that a submarine was sighted off the Maine coast. So the guns are trained upon every vessel they see. The raider is supposed to be a boat disguised as a tramp and to look much like the Mongolia.

This PM at 4:30 we all put on our life belts & were assigned to the boats. Six women were in ours #3 Miss Cohen, Miss Hoffman, Miss Alexander, Bess, Frances & I. Mac and Pawly are together in 11. We are glad we aren't all separated any way. There are about 3 men to every woman, besides the crew in the boats. It was quite interesting.

Tonight after dinner Pawly and I went up and walked the decks again. One of the sailor lads came & walked & talked & talked. He does enjoy telling us a lot of things but of course we don't know how true they are. Sailor yarns I suppose.

I certainly never thought that I'd be on a tramp steamer running a blockade. It sounds like some story rather than stupid every day life that I've usually had. But I certainly cannot complain that the last year has been stupid. It has been full of a little of many things and much of a few ????? But naturally I'll take what comes and be glad it's no worse.

5-24-17

Today has been very quiet. I've walked & read a book & watched the water and listened to the waves swishing against the ship's side.

This afternoon the boys boxed & we watched them & then they all came up on the deck & had potato races.

5-25-17

Today has been beautifully clear and quiet. Nothing happened except a card party this P.M.

This evening I watched the sunset and then the moonlight on the water. It is like a beautiful dream.

5-26-17

Today the sun shone on the white caps but the horizon was misty. I sat and read and dreamed as usual.

The sailor lads told me we passed an ice berg last night and also that they found a small fire down in the lower regions of the ship and the fog was so thick they could not see their way about on deck.

This evening we are again in fog and the waves are running the highest I've seen them. The spray reaches to the top deck and the ship rolls quite a lot.

The sailors prefer the fog as it is much safer as far as submarines are concerned and they don't have to watch so closely.

5-27-17 Sunday

I'm certainly getting lazy. We've nothing to do but sleep, eat & walk. Some of the girls have quite a time flirting, but I had a set back and have reformed. Went out in the bow this morning and talked to the sailors. They are on watch all the time even tho we are not yet in the danger zone. We do not enter that till about Wednesday.

We had very nice church services at 10:30 A.M. Some of the privates sing well and one plays the violin very well indeed.

It is misty in the distance and the sun shines only occasionally. We are certainly having a fine voyage.

5-28-17

A nice lazy day—foggy but not so that the sun can not get thru. The only event I can remember is washing handkerchiefs & sewing buttons on my coat.

Last night Elizabeth Cleveland & I went on the top deck to watch the moonlight on the water. The sailor lad Choice came up to talk to us. He is a funny boy and so perfectly good natured.

5-29-17

Another long, lazy, lazy day. We are to have our pictures taken the PM. We are to wear our Brassards with our uniforms. We passed two ships this morning but they looked like only a column of smoke on the horizon.

5-30-17

It began to rain at 11 o'clock so we all had to come down to the lounge or to the hurricane deck. I staid out almost all day and watched the waves and the rain. At noon we passed a British freighter. It was the first ship we had really seen since the day we left New York.

We enter the danger zone some time tonight. I do hope it is warmer and not so rough if we have to take to the life boats.

Miss Cleveland and I were nearly drowned by the spray from one big wave.

5-31-17

We've been one whole 24 hours in the war zone and so far have sighted nothing more than a tramp steamer, away off on the horizon. The waves have been quite high and as we zig zag every 10 minutes we roll about quite a lot. The water is beautifully blue and tonight the moon is glorious even tho dangerous to us. Elizabeth Cleveland & I went up on deck wrapped up in blankets and just looked and looked.

Frances, Mac, Pawly & Bess are out parading with the boys on lower deck.

Last night the sailors put up cots out on deck beside their guns. And some special precautions were taken to prevent lights showing when the doors are opened.

I'm getting into the card habit again but it is a social thing to do and goodness knows I haven't many social accomplishments.

Last night, Ellen Thomsen, E. Cleveland, Bess & I played 3 games of 500. Bess and I were beaten but it was very close.

Later—

Have been out on deck watching the waves. We've been in war zone since 9 P.M. but there is practically no danger as the waves are too high.

6-1-17

It was midday on the 1st of June
On the deep blue ocean
That the five blasts then the three
Set us all in motion.

Tis Ship Line No. 9 cried Howe
In his quiet manner
I'll get my life belt
Just in time to see her when they land her
Port side watch sub ahoy
Cried the Lookout Forward
Man your guns
Damn those Huns
Starboard Lifeboat lower

A belted fair one wrapped in cork
Remarks "We've got a whale!
But since it must be that He die
I hope it is a male!"

Then in his wrath the Captain bold
Proclaimed aloud "Oh, Pish!
I'll sail no more the war zone sea
If that blamed thing is a fish!"

However it may seem to you
We all agree to this
That whether whale or submarine
It was, alas, a miss.

Now, doctors all, you will admit
That in the point of thrills
Bagging a school of submarines
Is better than peddling pills.
—Dr. Marbury

About 12:20 pm, I was standing under the bridge watching the rain & clouds when I heard the man on the bridge shout "Port side abeam" and the sailors went tumbling to their guns around. I was so anxious to see that Submarine but just couldn't see it & then the danger whistle blew and one of the officers grabbed me by the arm & said "Life belts" and we ran down the steps & I didn't see anything. We all got our belts &

came back into the lounge and waited. Not many were frightened but some were excited. The guns went off and we all waited until the mate came in and said "It's gone and there's our convoy." But after the British gun boat had come up, she—the sub—came up again and was fired at. Her torpedoe went astern about 50 yards probably because of the zig zag course we were taking and also because of the rain. But it was a close call. We fired 5 shots but don't know if any were effective. The British convoy staid with us then racing back and forth across the bow like a puppy.

6-2-17

My but everyone was excited and kept watching all afternoon & evening and no one went to bed till terribly late and many not at all.

Miss C. and I sat on top deck after the Dr's entertainment & watched the moonlight & talked. We got to bed about 1 and then had the watch call us at 3:30 to see the sun rise. It was very beautiful.

And we could see the British boats coming out—counted eight against the horizon. We reached the harbor at Falmouth at 6:30. The hills are beautiful. Just like patchwork quilts—the green fields with hedges and then the shrubbery and the little houses nestling down among the trees. Pembroke castle was very plain from the mouth of the harbor. It is high on the hill above the village.

Afterwards we were towed around a headland and anchored outside Falmouth. We're now waiting to be inspected and taken ashore.

We hear that two boats were torpedoed or mined last night—one in front and one behind us. Evidently we have work to do or we wouldn't have escaped.

Evening, 10 o'clock.

At noon we watched the boys go on the launch & take their departure for Black Pool. Then at 2:30 the officers and nurses went ashore. We are put up at 3 different hotels. Ours is the smallest and is a delightful place. It is so clean and white & wholesome.

About 5 we walked up to Castle Pendennis & looked it over. It was built in 1542 and was quite interesting even tho it has some modern additions. Soldiers are quartered there now and it really is a fort. The privilege of seeing it was an honor. It has not been open to visitors since the war began. The walk up was beautiful and so sweet from the Hawthorne. Many wild flowers are everywhere but they aren't much like ours.

After dinner we drove through the better part of town & out into the country.

The hedges and the walls and the narrow lanes and the wild flowers & ivy and everything is certainly a treat. It is like a dream rather than reality. The houses are very picturesque and exclusive—mostly walled in—but their cottages or flats or whatever they call them are hideous. But we're all just wild about the place.

6-4-17

We came up to London by train yesterday. We left at 11:30 and were about 8 1/2 hours coming the 140 miles. The English country is beautiful but before we reached London I began to wish I could see something besides hedges and tiny fields and stone houses.

We saw Windsor Castle at a distance. It was quite imposing. We were sent up to the Hotel York in big busses.

We have heard that there was more than one submarine after us and what we had taken for whales were really torpedoes. We evidently were just fortunate, as everything ahead and behind us was torpedoed & sunk. This morning Elizabeth & I came over to the other part of the hotel for a shampoo. We first made the mistake of going to the "hair dresser" and found it to be a barber shop.

We aren't at all delighted with London so far but I think it's because we are tired. I was all in but my shoestrings last night.

We hear that the King and Queen have been receiving the various American units very informally and we hope that we also will see them.

Everyone has been delightfully friendly and we think English people are not all snobbish. Their accent is so funny but they seem to have more difficulty in understanding us than we them.

6-5-17

Yesterday PM we saw Madam Bassard's Wax figures and the Tower. I was very enthusiastic about the Tower. It was so full of historic interest.

But I do think London is an ugly place. We went to the American Embassy to sign up for passports and to have our pictures taken this morning. This afternoon we will go to the House of Parliament.

We have just learned that the Mongolia was reported sunk a week ago or more so our people have probably been notified. It is a mystery how such a thing was started.

6-6-17

We did not get into the House of Parliament as no one is admitted while they are in session except men, so we went across the way to West Minster Abbey & I surely did enjoy that.

In the evening we went to Queen's Theater to the Third Floor Back—admitted free on our uniforms. Didn't care especially for it.

This morning we saw the armed guard change at Whitehall—Knights of Pythias Parade—had pay day at 12 and this afternoon went to St. Paul's and to Kensington Gardens.

6-9-17

For three days I've had no time to write so I haven't. Day before yesterday we took our London guides in hand and went exploring. We saw lots of little things, such as the statue of Queen Elizabeth that used to be over the old Ludgate, the Old Curiosity Shop (not authentic), a house where David Garrick had lived, the Savoy Chapel (very interesting) and the law courts, The Temple with its interesting banqueting hall and funny old round church with the crusaders tombs. Then we found the old inn where Benjamin Johnson used to go, "The Cheshire Cheese." It is on one of those queer little courts off the Strand.

Then we went across London Bridge and saw the old inn where Charles Dickens used to stop. It is "Ye George's Inn." We visited the site of the old Tabard Inn and the old Southwark Church. Then we came home by the Tube which I loathe and had dinner and went to see Aida at Drury Lane.

Yesterday, nine of us went out to Ipswich to Stoke Park, Lady Henneker Heaton's country home, or rather, one of her homes. It was delightful. Lady Heaton was very pleasant and not at all snobbish. Seemed just a real person.

In the evening we went to His Majesty's Theater to "Chu Chin Chow." It was very beautifully staged.

Today we went shopping and to the House of Parliament and also to West Minster Abbey. Did I say before that the most precious things, such as the tombs of Queen Elizabeth, Mary Queen of Scots, and Edward the Confessor and Henry VII are protected from bombs by bags of sand. And the real coronation chair has been removed.

Tonight we are all tired and a trifle homesick. We are so sick of English cooking.

6-10-17

This morning we took a ride out to Richmond and back. This P.M. we went out to Sir Thos. Lipton's in four busses. We surely had a good time. Red, white & blue bouquets, American flags, lemonade, ice cream, speeches, music, readings, moving pictures & a box of candy each for the girls & cigarettes for the men. He certainly knows how to entertain. Tomorrow we start to France.

June 11, 1917 11 P.M.

We took the train at 3:15 pm and arrived at Folkstone at 5:15 went right onto the boat and crossed the Channel. We all had to wear life belts and we with a troop ship and a hospital ship were convoyed by a torpedoe boat and air ship, a "blimp."

We were put into big automobile busses and brot out to our happy home. We got here at 9 or 9:30 and were half starved.

The nurses fed us bully beef, bread, butter and coffee and then showed us our "huts." They are very nice but of course very bare. We have board beds with straw on top.

6-13-17

Yesterday we were assigned our wards. I drew one that usually has minor cases. Miss Behler is my assistant.

Our ward is practically empty tonight. I understand that we are to get no convoys till the hospital is almost empty. Then the English will depart and leave us in peace.

War is a dreadful thing. I only hope that it will not last long. I can't bear to think of our boys being mixed up in such frightfulness.

The English nurses are not keen about us and we are equally not so about most of their methods.

This was Laura's final diary entry. We don't know why she didn't keep it up afterwards—perhaps just too busy nursing. But it is significant that she saved it for the rest of her life.

2. BIOGRAPHICAL NOTES ON KEY PERSONS IN NARRATIVE

Major Frederick A. Besley, professor of surgery at Northwestern University Medical School from which he had graduated with an M.D. degree in 1894, was 48 years old in 1916 when he went to Washington, D.C., for his reserve commission and to organize Base Hospital 12.

Major Christopher C. Collins, Medical Corps, U. S. Army, the unit's initial commanding officer.

First Lt. Gerard N. Krost was born in 1888 at Crown Point, Indiana, earned a B.S. degree at University of Chicago in 1912 and his MD from the Rush Medical College in Chicago in 1914. He interned at St. Luke's Hospital in Chicago and joined the faculty of the NU School of Medicine in 1917. He was married to Reserve Nurse Carrie G. Krost, also of Base Hospital 12, and, in a most unusual fashion, the two served as husband-and-wife medical team partners for the entire overseas period.

Capt. Milton Mandel, born in 1881, earned his M.D. from Northwestern in 1904 and was affiliated with Mercy Hospital in Chicago at the time the base hospital was activated.

Capt. Walter H. Nadler was born in Peru, Indiana, in 1889, obtained his undergraduate degree from Northwestern in 1910 and his M.D. in 1913, and joined the Medical School faculty in 1916.

First Lt. M. Pinson Neal, M.D., Base Hospital 12's original pathologist.

Capt. Payson L. Nussbaum, born 1879 in Middlebury, Indiana, attended Indiana University before transferring to Northwestern, where he earned his undergraduate degree in 1905 and his M.D. in 1911. Trained as a gynecologist, he joined the NU Medical School faculty in 1915.

Capt. Kellogg Speed graduated from the University of Chicago where he had been captain of the 1900 football team. He obtained his M.D. degree from Rush Medical College and was a member of the NU Medical School faculty beginning in 1913.

Chief Nurse Daisy Dean Urch, born June 9, 1876 in Davisburg, Oakland County, Michigan, was selected by Dr. Besley to recruit nurses for Base Hospital 12. At the time she was the chief nurse, supervisor and assistant instructor at Illinois Training School for Nurses, serving Cook County Hospital, and a 1913 graduate of the school. As a consequence of her endeavors, 40 of the 62 original nurses in Unit 12 were ITS graduates. Earlier, Daisy graduated from the Kindergarten Department at Ferris Institute in western Michigan, spent two years at the Normal School (now Eastern Michigan University) and was a teacher and later the Principal at West Ward Public School, Munising, Michigan, before abandoning teaching for nursing in 1910. When called to active service, she (and many of the other nurses) lived at the nursing school, 509 Honore Street, adjacent to Cook County Hospital. Along with Reserve Nurse **Bertha Alexander**, she was one of two original Northwestern Unit nurses named in dispatches by British supreme commander Field Marshall Haig in December 1917.

When they enlisted as reserve nurses on February 21, 1917, Laura Huckleberry and seven other 1913 ITS graduates shared an apartment at 453 Belmont, near Chicago's near-north-side Lincoln Park. One of the roommates, Mary Lunquist, was rejected for health reasons. The other roommates who departed with Laura for France were:

Mrs. Edith Ayres, a 36-year-old widow from Attica, Ohio. Mrs. Ayres, born Clara Edith Work, a schoolteacher before getting married in 1903 to Wayland D. Ayres. She was widowed when her husband, a grocer in Attica, died of tetanus in 1906 as a result of a minor injury. She continued the grocery business and later worked in a dry goods store until leaving to attend ITS in 1910.

Bess B. Gambee, aged 32, also from Attica, Ohio, a small town in northwestern Ohio.

Frances B. Hampton from Frankfort and Catlettsburg, Kentucky, aged about 37 in 1917.

Emma Matzen, aged 37 when the unit embarked for Europe, whose home town was Columbus, Nebraska

Edith (Mac) McMillin, of Lafayette, Indiana

Ella R. (Pawly or Polly) Pawlisch, born in North Freedom, Wisconsin

May (Conny) Connard of Decatur, Illinois, born March 24, 1889, also a 1913 ITS graduate, who after other adventures with the Red Cross was in the first wave of BH 12 reinforcements in October 1917. She had been a roommate of Laura and other classmates at 453 Belmont in Chicago before undertaking a Red Cross assignment to Germany in 1916.

Another ITS graduate in the unit was **Elizabeth C. (Bess) Lyon**, of Eau Claire, Wisconsin, a member of the class of 1914. In 1975 she contributed a memoir about the unit's history to an ITS Alumni publication.

Laura's assigned roommate in France was **Elizabeth T. Cleveland**, born in Yorkville, Illinois, January 20, 1880, who lived at 31st and California Avenue, Chicago, when she enrolled in the Northwestern Unit. Her hometown was given as Decatur, Illinois. She was not an ITS graduate. Elizabeth Cleveland married Walter Morf in 1923. They lived at 3936 North Kildare in Chicago when she died April 15, 1931. She had been the secretary of the Chicago Red Cross Nursing Service.

Some other persons involved in this narrative:

According to *The Syllabus*, published by the Northwestern University Junior class in 1919, "nearly eighty Northwestern men responded to [Dr. Besley's] call, who, with the [University of] Chicago men and a few others, made up the unit." A 1919 Chicago newspaper story upon the unit's returned reported forty of the unit's 150-plus enlisted personnel were University of Chicago men. An Internet search revealed three former Knox College upperclassmen as BH 12 members. Two of the enlisted personnel stand out:

Private George R. Baker of Rock Island, Illinois, 25, a student at Chicago's YMCA College when he joined the Northwestern Unit in May 1917.

Private Arthur R. Hall, a favorite orderly who had been a law student at Northwestern when he enlisted.

The Northwestern Unit Chaplain:

The Rev. Jesse S. Dancey, pastor of Englewood First Methodist Episcopal Church, Englewood, Illinois.

Wounded British officers:

Lt. Duncan Darroch McCosh, born December 30, 1888, in Paris, Ontario, appointed to 32nd Battalion, Canadian Over-Seas Expeditionary Force, July 7, 1916, after nine months at Canadian Officers Training Corps. He was a bank clerk before enlisting.

Lt. Robert Murray, born May 19, 1883 in Hawich, Scotland. Enlisted November 9, 1914 in the 29th Battalion, Canadian Expeditionary Force, from Vancouver, British Columbia. Was wounded, decorated for bravery, and given a field promotion to lieutenant.

3. QUASI-PUBLIC ROLES OF PRIVATE ORGANIZATIONS

The relationship between American military and civilian organizations has gone full circle in the last century. Today, with the aim of stretching its budget, the U. S. Department of Defense is rapidly endeavoring to privatize all support elements for the military not trained and armed for combat, such as food providers, transportation, some communications and even medical personnel not deployed with fighting units.

Further, the inclusion of any religious symbols in the public arena, especially those of historically dominant Christianity in America, is the subject of continuing civil rights criticism and litigation at all levels.

But at the opening of the 20th Century, it was a very different story. According to the official Young Men's Christian Association history, in 1902, Congress authorized the erection of permanent YMCA facilities on military bases. Almost 26,000 YMCA staff and volunteers performed YMCA work during the First World War, some of it years before America entered the war.

With its more than 1,500 canteens and post exchanges, the YMCA fed and entertained more troops during World War I than did any other welfare organization, including the Knights of Columbus and the Salvation Army. It was common for Catholics and Jews to use Y buildings for religious services. In all, the YMCA performed more than 90 percent of the welfare work of the time, mostly in the form of running canteens and post exchanges.

4. AMERICAN RED CROSS AS U.S. ARMY MEDICAL CORPS RESERVE

A similar situation evolved for medical and nursing services. During the Spanish-American War of 1898, Dr. Anita Newcomb McGee, Vice President of the National Society of the Daughters of the American Revolution (DAR), was placed in charge of selecting graduate nurses for the Army. Military nursing had been almost dormant since the Civil War. Dr. McGee suggested that the DAR act as an application review board for military nursing services. Thus the DAR Hospital Corps was founded, with Dr. McGee as its director. Following the war, at the request of the Surgeon General, Dr. McGee wrote a bill to establish a Nurse Corps (female), enacted by Congress in 1901. In 1911, the enrolled nurses of the American Red Cross were designated as the primary source of reserve nurses for the Army. The federal budget included provisions for the private Red Cross in support of its governmental responsibilities.

Even before the U. S. entered the war on the side of the Allies, American volunteers were providing medical services to both sides of the conflict in Europe, American Field Service ambulance drivers, a Harvard hospital unit and Red Cross teams of doctors and nurses.

5. PROLOGUE: NURSE MAY CONNARD'S ADVENTURES

Nurse May Connard from Chicago recounted an unparalleled adventure in a 1977 memoir:

In 1916 I was asked to go to Germany with three doctors and one other nurse. World War I had started and groups had been organized to co-operate with the Red Cross in sending teams of doctors and nurses to set up hospitals in various countries in Europe on both sides of the war. One such group was to be from Chicago. The chief was a surgeon under whom I had worked in the operating room. It was also necessary to speak some German. The other nurse spoke German well but had little operating experience. The two younger doctors had just finished their internships. The chief's wife also went along. She was a nurse but never acted in that capacity.

We sailed from New York in February 1916 on a Scandinavian liner and went north to Halifax, where we remained in the harbor two days, but were not allowed to leave the ship. Then the ship's course took us north of the British Isles near Iceland, in order to avoid German submarines. The weather was stormy and cold, but I was lucky not to be sea-sick, and could walk the deck with the only other passenger who could come out, a Dane who was returning home with his wife and daughter, from representing Denmark at the San Francisco Fair. Later I visited their home near Copenhagen.

Due to the war we stopped in the harbor in the Orkney Island. I awoke in the morning to the sound of sea-gulls screaming. I'm sure there were millions of them. Again we were not allowed off the ship and were carefully examined for security reasons. I suppose there were some German spies on board, but they were not detected. During the rest of the voyage we sailed at nights without lights, constantly warning our lifebelts, and had frequent warnings of submarines sighted, or mines suspected.

We docked safely in Copenhagen. Our hotel reservations were non-existent due to the city being full of travelers trying to get away from Europe. We found a place to stay near a night-club where the noise and music lasted all the night. I longed for the quiet of the ship, with only fog-horns and sea-gulls to disturb one.

In the morning we boarded a train for Warnemunde, the border town. There we were separated for the Germans to examine us. They were even more afraid of spies than the English. I had to undress and my body was closely inspected, even my mouth and teeth. Each article of clothing was gone over carefully. The baggage was unpacked, tooth-paste squeezed out and cakes of soap cut in halves.

At last I was allowed to go and board the train for Berlin, which started almost immediately. I could find no trace of the others. The porters and other passengers were most concerned. They could understand my school-girl German a little better than I could their rapid speech. Finally I understood there were two sections to this train and I should get off at Rostock and wait for the next section.

I waited about an hour before another section pulled in. I had no idea how to get on. No porter appeared to open any doors. Luckily a fellow passenger of the ship saw my bewilderment, opened a door and pushed me in just as the train was starting. After going through several coaches I found my party. They were not at all concerned, expecting to find me at the Berlin Station, where I should have been except for so much advice.

In Berlin we stayed at a huge hotel. Our room had a bath big enough for another room. There was very little central heating, but we were getting used to cold weather, for none of the places seemed to be heated. Berlin was a beautiful city, with the Unter den Linden Ave. bordered by linden trees and magnificent statuary, including the Brandenburg Tor. The people were most friendly and anxious to please.

After a few days we were sent to the place where we were to be stationed. This was Naumberg-an-der-Saale, an old, old city. The streets were narrow and very crooked, paved with rough bricks, and no sidewalks. There was an old cathedral with a walled garden, built in the 12th century. In the center of the town was the usual square market-place, government buildings on one side and shops on the other. Several days each week, farmers brought their produce and had stalls to display. Among the shops was a coffee-shop where nearly everyone went for afternoon coffee and cakes. We stayed at the "Hotel zum Schwartzen Ross" (Hotel to the black horse). It was very old and not too comfortable, so we hunted other quarters as soon as possible. Miss Eagle (the other nurse) and I found rooms at Frau Berger's pension. We had a bedroom and a living room with a balcony. Frau Berger spoke only German, as did the maid, so I had to learn quickly. From time to time other roomers came, sometimes army people, and for a long time the woman who was x-Ray technician in our hospital./

A large old school building was turned over to us to fit up as a hospital. German workers, orderlies and nurses were assigned to us, some of whom spoke English. We had brought many supplies with us but it was hard to find what was lacking. In a short time we were ready for work. Beds had been set up in the rooms, and with the help of an orderly a big room was scrubbed and fitted up as an operating room. Our first patients came as a convoy from a field hospital. We were all glad to be working at last. Each time a convoy came in, the operating room was very busy for several days. Gun-shot wounds require surgery. Shrapnel must be removed, sometimes amputations of legs or arms. Each wound was badly infected, sometimes with gas-gangrene, so I often worked far into the night on those days. When we were caught up, there was only minor surgery and dressings until the next convoy.

Food was scarce, even at first, and we were given ration cards. As time went on, it became more and more scarce. Frau Berger had our cards. Each day the maid had to stand in line at the markets to get our share. Before we left we were limited to half a pound of meat a week, an egg a month, no milk or butter. Vegetables were grown in every available spot, including front yards. We practically lived on cabbage, potatoes, and onions. There was ersatz coffee, made of parched grain, and heavy dark bread. A few things such as fish and a sort of coffee-cake, made without sugar, were available at the cafes. No matter the taste they were gratefully eaten after one of our long walks.

Walking through the woods and along the river was one of our main pastimes, especially on Sunday afternoons. The river Saale was beautiful. To cross it there was a flat ferry attached to overhead wires. The driver just pulled it across by the wires. Once I went on an excursion boat where we stopped at an old ruined castle high up on a bluff.

During the slow seasons I tried to visit as much of Germany as possible. Most of the time I traveled alone, for one nurse had to stay at the hospital. We were not far from Weimar, Goethe's home, where there was a lovely theater. Several times I went to the opera there. I also went to the opera and shopping at Leipzig. The opera "Parsival" lasted all afternoon and evening; there was time out for a light supper. At intermission everyone went promenading in the foyer, drank beer or wine and ate whatever they could buy. Eventually I went to Munich, Berchtesgarten, Konigsee, Alt Rosenberg and Nurnburg.

When arriving at a place one had to report to the police before being allowed to stay the night. At Berchtesgarten I was taken to the home of the policeman who was awakened in order to sign the papers for me to stay at the hotel. At Berchtesgarten, with a guide and others, I climbed a mountain on foot staying at a mountain shelter high up in the clouds at night. I've often wondered if it was the same mountain where Hitler's hideout was.

Another time I went to Sassnitz-an Rugen, up on the Baltic sea. A nurse classmate, who had returned to Sweden and was working with the Swedish Red Cross, was located at a Prisoner Exchange Station there. Germans who had been taken prisoner by the Russians were exchanged for an equal number of Russians, who were brought in by a Swedish ferry. The Russians were examined by the Swedish before boarding the ferry and the Germans had been examined before they were allowed to come. This was to insure no communicable disease was present, and that no one who could again become fit for military duty would come. Red Cross workers were on hand to help the prisoners find their families. After the prisoners disembarked, we were entertained on the ferry. Much welcome cakes and real coffee were served.

As the situation grew worse, the sinking of the Lusitania brought an end to our usefulness in Germany. The people in Naumberg remained very friendly, as we prepared to leave after diplomatic relations were broken off, giving us farewell parties and gifts. We went to Berlin where we stayed several days, visiting the Spanish and Danish consuls to get our papers in order, so they would let us out of the country. Then we went on to Copenhagen, Denmark. Many travellers were there waiting to get back to the U. S. A., and hotel room was at a premium. Another team of nurses and doctors who had been in Austria were also there.

No passenger steamers were sailing, so there was nothing to do but wait and be happy and also see some of Denmark. I contacted the friends I had met on the voyage over and was invited to their delightful home. All the Americans in the hotel became very friendly.

While there I crossed by ferry to Sweden and spent a week in Stockholm, visiting my Swedish friend and her family.

After two months, we secured passage on a Norwegian steamer, which was running the blockade. We were not allowed to board at Copenhagen, but our baggage was loaded there and we went by train to Oslo, Norway, arriving the day before sailing.

Although we boarded the steamer in the evening, it did not leave until midnight and then without lights. By morning we were at Bergen, Norway, where we anchored until night. At that time German submarines were in the North Sea, trying to blockade Scandinavian countries. All ships had to run this blockade, watching out for U-boats and mines. We cross just below Iceland and arrived safely in New York, after two weeks on board. Everyone was surprised that we were home, for our letters had been delayed, and we'd left suddenly when the opportunity came.

6. ROSTER OF BASE HOSPITAL 12 OFFICERS AND NURSES, MAY 1917

From unidentified Chicago newspaper, Thursday, May 17, 1917:

...In addition to Dr. Besley, who has been commissioned a major in the regular army, Dr. Milton Mandel, assistant medical director, and Dr. Kellogg Speed, assisting surgeon, hold commissions as majors. Other commissioned physicians of the unit are:

CAPTAINS

Chase, Martin M, 23 E. Madison st.

Dale, Philip M, Granite City

Koch, Sumner L., 4505 Oakenwald av.

Lebovitz, Jos. J., 9154 Commercial av.

Nadler, Walter H., 30 North Michigan av.

Nusbaum, Payson L., 521 Oakwood blvd.

Sherman, Adam G., Aurora

LIEUTENANTS

Baker, Hillier L., county hospital

Barnett, Alex H., 3123 West 15th st.

Bird, John T., county hospital

Eaton, Robert M., county hospital

Glaspel, Cyril J., Grafton, N. D.

Harwood, William H., Joliet

Henkin, John, county hospital

Krost, Gerard N., 30 N. Michigan av.

Neal, Marens P., 2607 Prairie av.

Robertson, C. W., county hospital

Ryan, E. O., county hospital

Talbot, Edwin R., Mercy Hospital

Stranberg, W. L., 2460 W. 63d st.

Drs. Stanley W. Clark and Charles W. Freeman, dentists, also have been commissioned as lieutenants. The Rev. Jesse S. Dancey, pastor of the First Methodist Episcopal Church, accompanied the unit as chaplain.

Following are the names of the nurses who left Chicago last night for France with Red Cross hospital unit No. 12:

Alexander, Bertha '10, Thomsen, Ill.

Applegren, Emma, 3012 Blair place

Ayres, Edith '13, Attica, O.

Baker, Florence Edna '10, 4065 Ellis avenue

Behler, Clara '14, Lafayette, Ind.

Burcham, Daisy '12, Wellington, Kas.

Carruthers, Isabelle, 867 N. Dearborn street

Chapman, Harriet '10, Chicago

Cleveland, Elizabeth (Chicago and Decatur, Ill.)

Cohen, Rebecca '11, Chicago

Collins, Anna M. '14, Solon Springs, Wis.

Crawford, Ada L., Evanston, Ill.

Dean, Myrtle, Atlantic, Ia.

Deuel, Sarah M., Evanston, Ill.

Gambee, Bessie '13, Attica, O.

Greenwood, Elizabeth, Duluth, Minn.

Hampton, Frances '13, Frankfort, Ky.

Hinton, Florence '14, 509 Honore street (Decatur, Ill.)

Hoffman, Clara, 6752 Merrill avenue

Hostman, Louise '09, Chandlerville, Ill.

Huckleberry, Laura '13, North Vernon, Ind.

Jacques, Albina M. '16, Lake Linden, Mich.

Jensen, Aileen, Chebanse, Ill.

Jones, Margaret B. '15, Wellsville, Utah

Judy, Stella Maude '15, Sibley, Ill.

Kandel, Ernestine, Oconto, Wis.

Krost, Carrie G., West Salem, Wis.

Larson, Freda W. '15, Waupaca, Wis.

Lyon, Elizabeth C. '14, Eau Claire, Wis.

Lyon, Emily R. '12, Marysville, Mich.

Mahony, Katherine '12, Melrose Park, Ill.

Matzen, Emma '13, Columbus, Neb.

McDonald, Helen, Evanston hospital

McMillan, Edith '13, 3649 Crawford avenue

Miller, Lena '11, Dowagiac, Mich.

Morrisey, May, Plainfield, Ill.

Murray, Edith Maude '10, Ludington, Mich.

Paulson, Belletta '14, Granton, Wis.

Pawlish, Ella '13, North Freedom, Wis.

Pepoon, Lucile, 3842 Byron street

Pfaff, Helen I. '13, Marietta, O.

Poole, Frances M., Evanston

Powers, Margaret M. '07, Marinette, Wis.

Prouty, Bess L., Evanston

Purdum, Sarah E. '10, Shabbona, Ill.

Radcliffe, Alice M., Evanston, Ill.

Rettke, Minnie, 3414 Elaine place

Richter, Theda, Evanston, Ill.

Ruden, Clara D. '16, La Crosse, Wis.

Silcox, Edna '12, Frome, Ont.

Sparks, Helena, Aurora, Ill.

Spencer, Ruth '11, Ludington, Mich.

Stahl, Nellie M. '14, Burt, Ia.

Streitmatter, Budy '12, Lewis Institute, Chicago

Sweet, Olive B. '12, Iron River, Mich.

Thompson, Pearl W., Evanston

Thomsen, Ellen '13, Lewis Institute

Umberger, Grace '09, Hymer, Kas.

Urch, Daisy D. '13, Davidsburg, Mich.

Van Alstine, Annis '14, Gilmore City, Ia.

Van Ark, Bessie, Evanston, Ill

Wilson, Julia, Vinton, Ia.

Wood, Helen B., Evanston, Ill.

The editor has appended the graduating class of those nurses from Illinois Training School for Nurses, indicated after their names.

7. ROSTER OF NURSE REINFORCEMENTS, OCTOBER 1917

In October 1917, Base Hospital 12 was reinforced by the arrival of 9 additional doctors, 33 nurses and 75 enlisted men. The roster of additional nurses included:

Allport, Jennie, Chicago, Chicago Union Hospital

Anderson, Jane '17, Butler, Ill., Wesley Hospital,

Auty, Hilda W.

Bartlett, Edith T.

Connard, May '13, Decatur, Ill., Illinois Training School

Crawford, Catherine M.

Duggan, Elizabeth V., Chicago, Ravenswood Hospital

Dunham, Carolyn J.

Fuller, Bessie, Galesburg, Ill., Wesley Hospital

Grundy, Phoebe M. '10, Oshkosh, Wisc., Illinois Training School

Habeger, Ida C., Chicago, Garfield Park Training School

Ham, Zola P., Chicago, Peoples Hospital

Higley, Mabel A., Augusta, Mich., Grace Hospital

Horowitz, Gabrielle, Chicago, Garfield Park Training School

Hoskyn, Emma J. '15, Calgary, Alberta, Can., Grace Hospital

Jones, Elizabeth A., Oak Park, Ill., Wesley Hospital

Lambert, Gertrude A.,

McCall, Emma B., Washington, Iowa, Wesley Hospital

MacEwan, Marion J.

Mitchell, Lucy

Olson, Anna

Reagles, Vernie G. (Mrs.) '16, Nampa, Idaho, Ill. Training School

Rose, Mary A., St. Louis, Mo., German Hospital

Sherman, Frances J.

Simmonds, Violet S.

Smith, Sarah H., Chicago, Wesley Hospital

Van Winkle, Helena, Abingdon, Ill., Wesley Hospital

Weir, Mary A.

Welch, Grace C., Wadsworth, Ill., Wesley Hospital

West, Mary K., Fairfield, Iowa, Wesley Hospital

Wilson, Viola

Wood, Mary H., Gaffney, S.C., Philadelphia General Hospital, Pa.

8. ROSTER OF BH 12 COMMAND, APRIL 2, 1919

(from a newspaper clipping)

MAJORS

Payson Nussbaum, Evanston

Kellogg Speed, 122 S. Michigan av.

CAPTAINS

Joseph Lebowitz, 8448 Green Bay av.

Walter Strauberg, 822 E. 48th st.

Hillier Baker, 1711 W. Garfield blvd.

Alexander Barnett, 4939, N. Kedzie av.

Stanley Clark, 6806 Cornell av.

LIEUTENANTS

Walter Siewerth, 3606 N. Racine av.

Elmer Roberts, Oak Park

9. ILLINOIS TRAINING SCHOOL FOR NURSES

The Illinois Training School for Nurses (ITS) in Chicago was founded in 1881 by pioneering woman physician Sarah Hackett Stevenson with a contract to supply professional nursing services to Cook County Hospital and provide nursing training. It was modeled after Florence Nightingale's nurse training institution in London, and would break ground as the first such facility in the Midwest—three similar institutions already having been started in the East.

At the time, modern scientific medicine was still in its infancy. The germ theory was little understood; the cause of cholera, for instance, was not verified until 1889 in Hamburg, Germany. Many fundamental new chemical pharmaceuticals such as aspirin were just coming forth. Medical school education for physicians was to undergo fundamental upgrading and professionalism right after the turn of the 20th century.

The nursing training provided by ITS covered a three-year co-op program, in which students both learned in classrooms at the school at 509 Honore Street—where they also lived—and nursed patients at nearby Cook County Hospital. Normally students had to be high school graduates, and some even held college degrees. Many had been school teachers, which required only a high school diploma, before attending ITS.

The nursing school fell on hard times financially in the 1920s, and was taken over by the University of Chicago in 1929, which agreed thereafter to award B.S. degrees to the nursing school graduates.

10. AMERICAN RED CROSS INSTRUCTIONS FOR NURSES

INSTRUCTIONS FOR NURSES, 1917

Red Cross Nurses assigned to duty with military establishments in the United States are required to wear the white uniform of the Army or Navy Nurse Corps, with attached soft white collar which should conform to Butterick pattern No. 9530. Indian head, cannon cloth, or poplin may be used. Nurses wishing to purchase uniforms ready-made may do so from any of the following firms for about $3.50 each, express prepaid: B. Altman and Company, 34th Street and Fifth Avenue, New York City; Carson, Pirie, Scott and Company, corner State and Madison Streets, Chicago, Illinois; and J. Magnin and Company, Grant Avenue and Geary Streets, San Francisco, California. They may be secured in sizes 32 inches to 46 inches bust measure, and are known as No. 400 nurse's uniform.

The gray uniforms which are now being used for foreign service will not, for the present, be issued for service in the United States either with the Army or Navy, but will be used by Public Health Nurses maintained by the Red Cross in Sanitary Zones.

If nurses already have white uniforms which conform in essential points, general appearance, and specifications, to the uniform of the Army and Navy Nurse Corps, they may be used. Collars known as No. 1181, made by E. W. Marvin and Company, Fifth Avenue and Federal Street, Troy, New York, should be worn with these uniforms. They are made in one size, but as they are cut low can usually be adjusted to fit without difficulty. They will be sold direct to Red Cross nurses ordering them, at $15 a gross or $1.35 a dozen. Orders for less than one dozen will not be filled. A post office money order for the price of the collars, plus the probable amount of postage, should be sent. Money orders are desirable as ordinary checks may cause delay. These collars may also be secured at the Headquarters of the Atlantic Division, 44 East Twenty-third Street, New York City.

EQUIPMENT FOR FOREIGN SERVICE

Provided without cost by the Red Cross to nurses assigned to foreign service with Army, Navy, or Red Cross:

Hat

Outdoor uniform (blue serge)

Coat or heavy ulster

Black woolen tights, two pairs

Cape (if dress is used)

Gloves (tan kid)

Gloves (woolen)

Shirtwaists (two, white)

Shirtwaist (blue flannel)

Sweater

Gray wash uniforms, four

Aprons, eight

Cuffs and collars

Caps, four

Sleeve links, two pairs

Insignia one set (military duty)

U. S. Letters, one set (military duty)

Steamer blanket

Sleeping bag

Poncho or rubber sheet

Blanket roll

Raincoat

Rain hat

Moccasins

Nurses will require warm underwear and pajamas. As shoes and stockings are difficult to secure in Europe, it is advisable to take a generous supply of both, including heavy tan shoes. Two dozen pairs of stockings, several of which should be wool, would be none too many.

Arrangements have been made with dealers to supply these articles at greatly reduced prices; they may be secured after arrival of nurses in New York, through our representative at 44 East Twenty-third Street. In case a nurse does not report in New York City, special arrangements will be made to cover the situation.

The equipment issued may be modified at the discretion of the Red Cross to suit conditions under which the nurse is working.

WHITE UNIFORMS AND OTHER ARTICLES FOR NURSES ASSIGNED TO FOREIGN SERVICE

Each nurse should take with her two or three white regulation uniforms for dress occasions. It is suggested that she provide herself with bed socks; also a small alcohol stove sufficiently large to prepare hot drinks. A small United States flag, water proof belt, identification tag, collapsible basin, fountain pen and ink tablets, solid alcohol, and a small supply of concentrated foods, including chocolate, will be found valuable.

NOTE. Experience has shown that each nurse should have about $50 with her upon arrival in New York City. This also applies to dietitians and clerical workers.

BAGGAGE

Nurses detailed for foreign service may take to the point of mobilization one steamer trunk not to exceed thirty-six inches in length, one suit case or large satchel, or equivalent, and one blanket roll (the last named provided in New York). Trunks, suit cases, and blanket rolls must be marked conspicuously in paint as follows: Name.........…………..

No. of Unit…………………(if a member of such). Army Nurse Corps or Navy Nurse Corps (if a member). Loss and delay will be avoided if this detail is observed.

Nurses should carry in a light-weight suitcase or handbag a complete uniform so that they may be available for service should there be delay in the delivery of baggage.

ADDITIONAL EQUIPMENT

Required for service, United States or Foreign.

Extra shoes, high, black, with low rubber heels and broad soles (pumps and French heels prohibited for foreign or home service)

Rubbers, raincoat, and umbrella (raincoat is included in equipment issued free for foreign service)

Underwear in sufficient quantities (crepe suggested for summer; knit for winter)

Dressing gown

Bedroom slippers

Extra corsets

Stockings, generous supply

Dr. Johnson's foot soap, one cake

Box of corn plaster or adhesive plaster

Cold cream and other necessary toilet articles

Box of soap

Hypodermic set, clinical thermometer, and pair of bandage scissors

Watch, with second hand

If glasses are worn, an additional pair should be taken

Hot water bag

RETURN OF EQUIPMENT

|Nurses will be held responsible for all equipment issued to them by the Red Cross, and will be required to return the same on the completion of their service to Red Cross Headquarters, Washington, D. C. They should not fail to attach their names to articles returned.

11. THE S. S. MONGOLIA

The 616-foot-long *SS Mongolia*, which carried Base Hospital 12 from New York to Falmouth, England, in May 1917, was built in Camden, New Jersey, in 1906 by the New York Shipbuilding Corporation for Pacific Ocean passenger-and-freight service. The customer was the Pacific Mail Line which used her specifically for San Francisco-Hongkong trade. She was rated at a maximum speed of 16 knots and equipped to carry 1,818 passengers. When converted to a troopship in 1917, she was designated the *USS Mongolia.*

After World War I, the *Mongolia* returned to civilian service like the military units she had carried, first for the American Line between New York and Hamburg and then, in 1923, for the Panama Pacific Line for New York to San Francisco voyages. In 1929, she was sold to the Dollar Line of San Francisco for "round the world" service and renamed the *SS President Fillmore.* When the Depression hit, she was laid up in New York from 1931 to 1940, at which time she was sold to Cia Transatlantica Centroamericana of Panama and again renamed as the *SS Panamanian.* Once again she managed to avoid German torpedoes during World War II, but was scrapped in 1946 at Hongkong.

12. WARTIME POETRY AND VERSE

It is generally recognized in literary circles that the most famous—and touching—verse to emerge from the Great War was *In Flanders Fields,* penned in May 1915 by Major (later Lieutenant Colonel) John McCrae, a surgeon with the Canadian Army Field Artillery:

In Flanders fields the poppies blow
Between the crosses, row on row
That mark our place; and in the sky
The larks, still bravely singing, fly
Scarce heard amid the guns below.

We are the Dead. Short days ago
We lived, felt dawn, saw sunset glow,
Loved, and were loved, and now we lie
In Flanders fields.

Take up our quarrel with the foe:
To you from failing hands we throw
The torch; be yours to hold it high.
If ye break faith with us who die
We shall not sleep, though poppies grow
In Flanders fields.

In 1977, Dr. C. J. Glaspel of Grafton, North Dakota, the last surviving M.D. from the Northwestern Unit, sat down in his den to write "for the BH 12 Archives" his account of Dr. McCrae's funeral. Dr. Glaspel, who among other honors had been the Chief Medical Examiner for the state of North Dakota, said that McCrae had suffered all his life from asthma, and in the winter of 1918, became seriously ill with pneumonia, dying late in January.

"On February 1, 1918, I attended the funeral of this gallant soldier as an aid to Colonel Besley of Hospital 12. For the first time I saw what is now routine, namely Col. McCrae's horse, Bowfire, led by a groom and preceding the casket, empty boots in the stirrups in a reversed position, followed by the casket bearers and firing squad. Then at least 100 nurses in white veils and caps; followed by the mourners and the rank and file of all officers. It was a springtime day, one that was so lovely and I recall so well how the chaplain stated 'so we may leave him on this sunny slope facing the sunset and the sea.' This funeral was held in Wimereau, just north of Boulogne."

A researcher and editor in the 21st century, delving into the letters and memoirs of Base Hospital 12 personnel from 1917-1919 cannot help but note the widespread joy then in reading and writing verse, something hard to imagine in today's atmosphere of harsh email messages. Doctors, nurses, privates and patients all wrote verse. Several of these verses have been included in the narrative. It seems appropriate, then, to close this story with an anonymous verse found in the Base Hospital 12 file at the National Archives:

Erle and Laura, Easter 1957, Sarasota, Florida.

ETAPLES MILITARY CEMETERY

The sandy slope sinks westward to the sea;
The cattle feed along the verdant lea;
The fishing fleet has loosened from the quay,
And down the Canche files on it seaward way;
Above the woods that lie beyond the stream
Two light-house columns in their whiteness gleam
In pleasant contrast with the tinted tile
Of Paris Plage, which shares the morning smile
Beyond it all to the horizon line
Rolls the vast waste of the Atlantic brine.

This is the scene which lights the eye,
Raised from the graves that all about us lie,
For in this slope ten thousand solders sleep;
And for each one how many hearts that weep!
From this sad spot grief's mystic cords far reach
To every land that knows the English speech;
There are dull hearths by all the Seven Seas
And souls that struggle with Death's mysteries;
No treasure fleet shall now make glad the wave
To bring this wealth here wasted in the grave!

This is the lower view. But lift the eyes!
Behold the vision glowing in the skies!
The seed decays. Above it springs the flower
To clothe with beauty earth's remotest hour,
Distilling sweetness from the earthy clod,
A grateful fragrance wafted up to God;
While precious petals falling on the tide, Will ever bless the farthest ocean side,
Favored the house to which has come the chance
To plant its seed in such a soil as France.

ABOUT THE AUTHOR/EDITOR

MICHAEL W. R. DAVIS is a veteran journalist, author, historian, and retired public relations executive. He holds a BA in history from Yale University and an MS in Historic Preservation from Eastern Michigan University where he also served as an adjunct professor. He completed the course work for a doctorate in history at Wayne State University, Detroit.

After retiring from the Ford Motor Company Public Relations staff, he was communications director of The Evening News Association (ENA), a newspaper publishing and broadcasting company; and subsequently executive director of the Detroit Historical Society. As a journalist, Davis has specialized in reporting on business—especially the auto industry—and military history for consumer and trade magazines.

He was editor of *The Florida Business Letter,* a staff writer for *The Miami Daily News,*Miami correspondent for *Business Week* and the *Kiplinger Florida Letter,* and contributor to *Time* and *Life* before becoming assistant manager of the *Business Week* Detroit bureau.

Since 1990, Davis has been author, co-author, editor or a major contributor of 17 previous non-fiction books. He is a member of the Society of Automotive Historians and numerous local historical organizations, and a former elected director of the Defense Orientation Conference Association, the Pioneer American Society/Association for the Preservation of Artifacts, the Metro Detroit Chapter of the Society of Professional Journalists, and the Board of the National Automotive History Collection at the Detroit Public Library.

He was briefly a member of the Kentucky Army National Guard between Louisville Male High School and matriculation at Yale. He has lived in the same house in Royal Oak, Michigan, with his wife Karen for more than a quarter century.

Made in the USA
Las Vegas, NV
29 March 2026

44471031R00159